Geology and Mineral Resources of the Santa Catalina Mountains, Southeastern Arizona

A Cross-Sectional Approach

Eric R. Force

with sections by
Daniel M. Unruh and Robert J. Kamilli

Monographs in Mineral Resource Science No. 1
Center for Mineral Resources, The University of Arizona®
and the U.S. Geological Survey

Center for Mineral Resources
Gould-Simpson Building
The University of Arizona
Tucson, Arizona 85721

Published 1997
Printed in the United States of America

Library of Congress Cataloging-in-Publication Data

Force, Eric R.
Geology and mineral resources of the Santa Catalina Mountains, southeastern Arizona: a cross-sectional approach / Eric R. Force ; with sections by Daniel M. Unruh and Robert J. Kamilli.
p. cm. — (Monographs in mineral resource science ; no. 1)
Includes bibliographical references.
ISBN (paper): 0-9661233-1-X
1. Geology—Arizona—Santa Catalina Mountains. 2. Mines and mineral resources—Arizona—Santa Catalina Mountains.
I. Title. II. Series.
QE86.S42F67 1997
557.91'77—dc21 97-31863
CIP

MONOGRAPHS IN MINERAL RESOURCE SCIENCE

Text format: Joanne M. Newcomb
Plate 1 format: Susan Tingley

The Center for Mineral Resources is a joint venture of the Geoscience Department of The University of Arizona and the U.S. Geological Survey, for the purpose of furthering collaborative research in mineral resource science. To this end, its Monograph Series publishes extended original studies of topics in the scientific and societal aspects of mineral resources. The scope is international. The goal is to provide a venue for high-quality, timely, and affordable presentation, either of single-component studies or multi-component studies linked by a common theme, including color maps. Submissions should be entirely digitial and with documented previous reviews. More information can be obtained from the Center for Mineral Resources, The University of Arizona, Gould-Simpson Building, Tucson, Arizona, 85721.

Distributed by The University of Arizona Press, 1230 N. Park Avenue, Suite 102, Tucson, Arizona 85719.

Printed on acid-free paper.

Contents

Plates

Figures

Tables

Preface

This volume introduces Monographs in Mineral Resource Science through the Center for Mineral Resources, The University of Arizona, in cooperation with the United States Geological Survey. This series is intended to fill a need for publication of extended original studies of topics in the scientific and societal aspects of mineral resources. Although this first volume focuses on a topic close to Tucson, the intended scope of contributions is international. The goal is to provide a venue for high-quality, timely, and affordable presentation of results that do not readily find a home in existing media. As such, we envision intermittent publication of single or multi-component studies linked by a common theme. An important component will be publication of color maps and digital distribution of information

Mark D. Barton
Director, Center for Mineral Resources

Geology and mineral resources of the Santa Catalina Mountains, southeastern Arizona: a cross-sectional approach

Eric R. Force

Abstract

The study area, a classic metamorphic core complex, actually shows the characteristic features of such complexes only at the southern end of the range and grades into other rock assemblages to the north. The area thus affords the opportunity of analyzing the definitive components of a core complex separately. It measures about 48 by 11 km, traversing the central part of the Santa Catalina Mountains from the Tucson basin to the San Pedro basin. The dimensions of study lend themselves to a cross-sectional approach.

A supracrustal sequence 1.1 km thick consists of Middle Proterozoic Apache Group and Paleozoic rocks unconformably overlying mostly-granitic basement rocks. Units of special interest include voluminous pyroclastic rocks in the Pioneer Formation, glacio-marine rocks of Upper Proterozoic age, and ?Devonian fluvial deposits in a paleochannel on Mt. Lemmon.

Laramide intrusions of roughly granodioritic composition are mostly concordant, typically inflate the supracrustal sequence about 300 m, and define three mutually exclusive large domains. The two domains characterized by Leatherwood and San Manuel intrusions show impressive heating and alteration effects, whereas the central domain characterized by Rice Peak intrusions show few such effects due apparently to multiple small intrusions.

Eocene peraluminous leucogranite forms seven sills toward the southern end of the study area, for a total thickness of 4 km. An upper sill splits open the unconformity on basement rocks, and has a thick pegmatitic cap. Two mid-Tertiary granitoid plutons may have contributed to the heat budget of core-complex uplift.

Two Laramide terranes are present in the main range. Each has its own characteristic Laramide intrusions and hydrothermal alteration character. The northern terrane contains a basement-cored uplift and a possible foreland basin, delineated by monoclines and by high-angle reverse faults, and laccolithic Rice Peak porphyry.

In contrast, the southern terrane, best represented near the range crest, shows Laramide metamorphism, penetrative deformation and recumbent sheath folding, thrust faults, and concomitant crustal thickening. Such features apparently extended southward to include the metamorphic core complex area, where they have been largely obliterated by mid-Tertiary mylonitization. Laramide paleodepths as much as about 6 km are preserved; this figure coupled with metamorphism suggests high Laramide geothermal gradients in the range-crest area.

Crustal thickening due to Laramide deformation is only about 1 km, much less than that due to Eocene leucogranites. Both factors show suitable distributions to contribute to a crustal root in the area of core-complex uplift.

The Geesaman fault originated as a steep left-oblique Laramide fault with about 7 km of offset, but was then deformed. The Mogul fault was originally similar and parallel, but was re-activated in the late Tertiary.

The main belt of mid-Tertiary mylonitic rocks passes under the main range, whereas the detachment fault passes over it, preserved in only one inter-range part of the study area. The main mylonitic belt is about 600 m thick, but passes down into wholly-ductile metamorphic rock. Pure shear was a component of early mylonitic deformation. Mylonitic fabric is parallel to Eocene leucogranite sills; both are anticlinally folded in the forerange at the south end of the study area. Reconstruction of the depth of mylonitization suggests about 5 to 10 km.

Mylonitization of two ages, pre- and post-26 Ma, can be discriminated; the younger formed a partly-new trajectory after anticlinal folding of the older belt. Each phase of mylonitization can be tentatively correlated with a phase of detachment faulting in the originally shallower northern end of the study area.

Laramide mineral deposits of the study area include porphyry and skarn deposits mostly of copper. Less important mid-Tertiary deposits are fault-related, mostly of gold and tungsten. The distribution of deposits shows a relation to paleodepth at the time of formation.

INTRODUCTION

The Santa Catalina Mountains, along with the adjacent Rincon and Tortolita Mountains, were among the first to be described as "metamorphic core complexes" (Davis and Coney, 1979), and together form the apparent southeast end of a chain of such complexes across Arizona (fig. 1). By now, however, geologic knowledge of the Catalinas has fallen far behind the pivotal role they play in core-complex discussions. A comprehensive re-examination is needed.

This study was designed to address both areal and cross-sectional aspects of the geology of the Catalinas (fig. 2). The study area is quite long (48 km) relative to its width (10 to 12 km), and was chosen to take advantage of access provided by the Catalina Highway and the Mt. Lemmon-to-Oracle "Control Road" (fig. 3). Thus it traverses the eastern parts of the range. Parts of eight 1:24,000 quadrangles are included (fig. 3). The area extends from mylonitic rocks on the south to upper-plate synextensional deposits on the north, by way of two contrasting terranes of Laramide (Late Cretaceous to early Eocene) age.

PHYSIOGRAPHY

The Catalinas are Basin and Range mountains, separated in part from adjacent ranges by fault-bounded basins. The Catalinas differ from typical ranges in southern Arizona in their larger area, broader crest, and greater elevation; they rise to an elevation of 9157 feet (2792 m) at Mt. Lemmon (within the study area) above

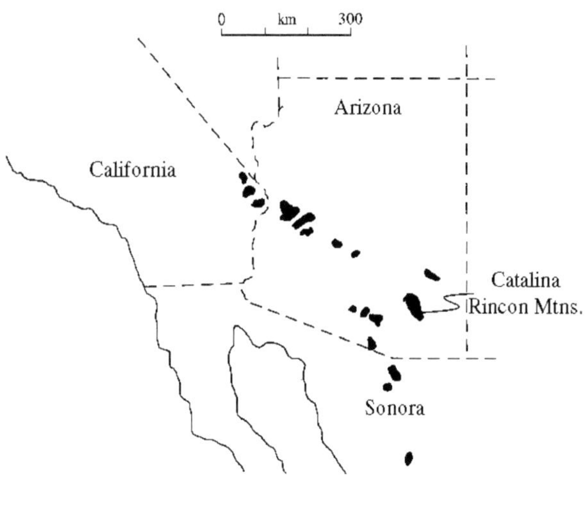

Figure 1 Relation of the Santa Catalina Mountains to core complexes of Arizona and California, from Coney (1980).

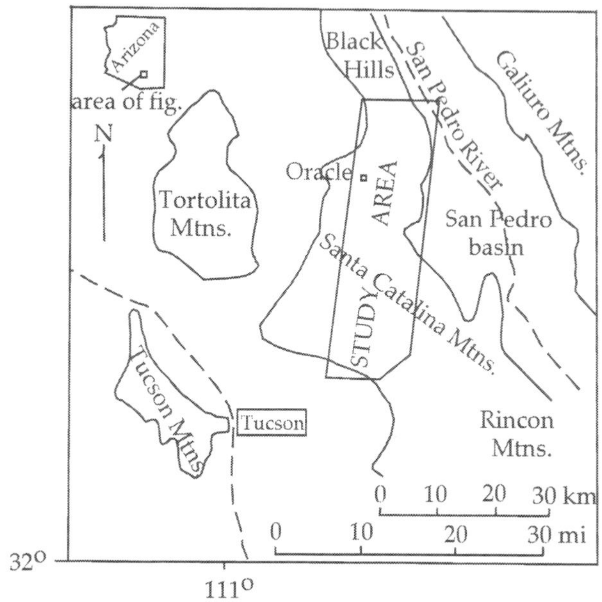

Figure 2 Location map showing the study area transect relative to nearby settlements and physiographic features.

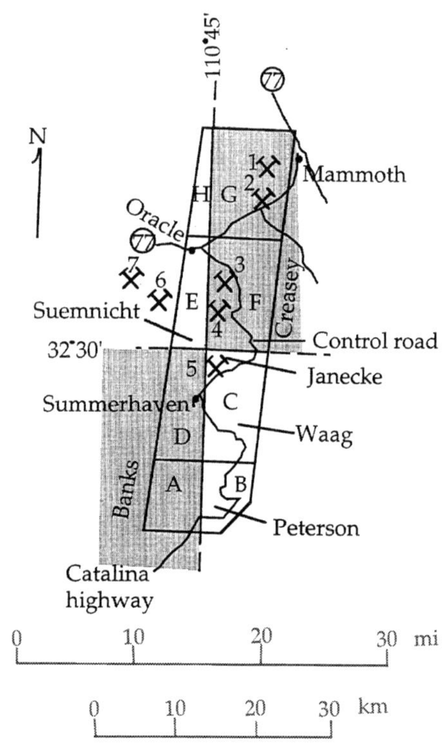

Figure 3 Index map showing quadrangles, previous mapping, and main mining districts. Quadrangles (7 1/2'); A. Sabino Canyon, B. Agua Caliente Hill, C. Mount Bigelow, D. Mt. Lemmon, E. Oracle, F. Campo Bonito, G. Mammoth, H. North of Oracle. Districts, 1. Mammoth, 2. San Manuel, 3. American Flag, 4. Campo Bonito-Southern Belle, 5. Marble Peak, 6. Burney, 7. Little Hills. Authors of shaded maps are Banks (1974) and Creasey (1965, 1967). Other maps (boundaries not shown) excluding superseded maps are Suemnicht (1977), Janecke (1986), Waag (1968), and Peterson (1968).

Figure 4 Photograph of the south-facing escarpment of the main range of the Catalina Mountains, showing palisades of slightly mylonitic Wilderness granite, from near Windy Point.

the Tucson basin to the southwest and the San Pedro basin the northeast, both with elevations locally less than 2500 feet (760 m). Within the study area, the crest of the range is mostly between 7500 and 8500 feet in elevation and trends roughly WNW - ESE, at an angle to Basin and Range trends. Rather steep escarpments face both directions from the crestal area (fig. 4), with slopes of 30° or greater widespread on some geologic units. On the Tucson (southwest) side of the range, an area commonly called the forerange is separated from the main range by alined valleys (fig. 5). On the northeast side, Oracle Ridge projects northward from the range; a broad plateau near the town of Oracle connects this ridge via the Black Hills and Tortilla Mountains to the transition zone of the Colorado Plateau. Thus the Catalinas form the southern end of an exposed oblique transect into undeformed crust, and this study treats them in that context.

Figure 5 Photograph looking along the alined valleys occupied by the Romero Pass fault zone in Sabino basin, from the western edge of the map area.

GEOLOGIC SETTING

The Catalinas are separated from adjacent basins and ranges mostly by extensional normal and detachment faults, and mylonites of extensional origin are the best-described aspect of their geology (Davis, 1980). In recent studies (Spencer and Reynolds, 1989; Dickinson, 1991; and others), the Catalinas have been considered the deeper parts of a top-to-the-southwest extensional domain of mid-Tertiary age that surfaces in the western part of the Galiuro Mountains (fig. 6). The Pinaleno Mountains northeast of the Galiuros are said to record similar extension in the opposite direction. Deeper-level extension was originally toward the southern end of the study area, and shallower-level extension toward the

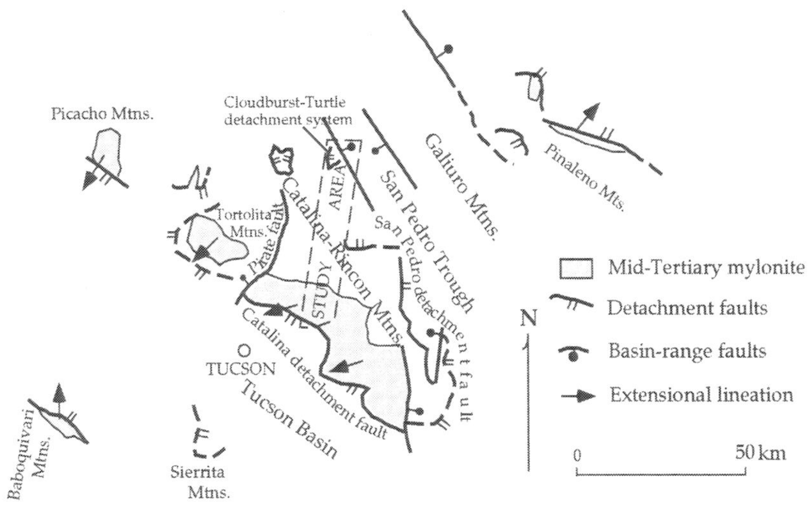

Figure 6 Map showing the study area relative to basin and range structure, mylonitic core complexes, detachment faults, and mid-Tertiary extension directions; modified from Dickinson (1991) and Spencer and Reynolds (1989).

north. Only the southern part of the Catalinas shows the mid-Tertiary semi-ductile extensional fabrics characteristic of metamorphic core complexes of the region.

Rock sequences in the Catalinas are a rather thin pile of Middle Proterozoic through Tertiary supracrustal rocks, typical for the region, unconformably overlying Early and Middle Proterozoic granite and schist (fig. 7). This sequence was inflated by intrusion of sills of Proterozoic, Laramide, and Eocene age (fig. 8), and thickened by thrusting in the central parts of the study area (fig. 9). The intra-Proterozoic unconformity was split open by the intrusion of Eocene leucogranite sills thousands of meters thick. In the mid-Tertiary, the structurally lower rocks now forming the southern end of the range were mylonitized during extension, and the entire pile tilted to the northeast as the mylonites arched upward. In the forerange, these arched mylonitic rocks control the shape of the land surface (fig. 10). Similar mylonitic rocks also define adjacent metamorphic core complexes.

Shallower-level extension occurred along detachment faults that ring the Catalina, Rincon and Tortolita Mountains (fig. 6). Detachment faults and other low-angle normal faults are exposed in the northern, east-central, and southern parts of the study area, but originally must have passed over the Catalina range, based on regional relations.

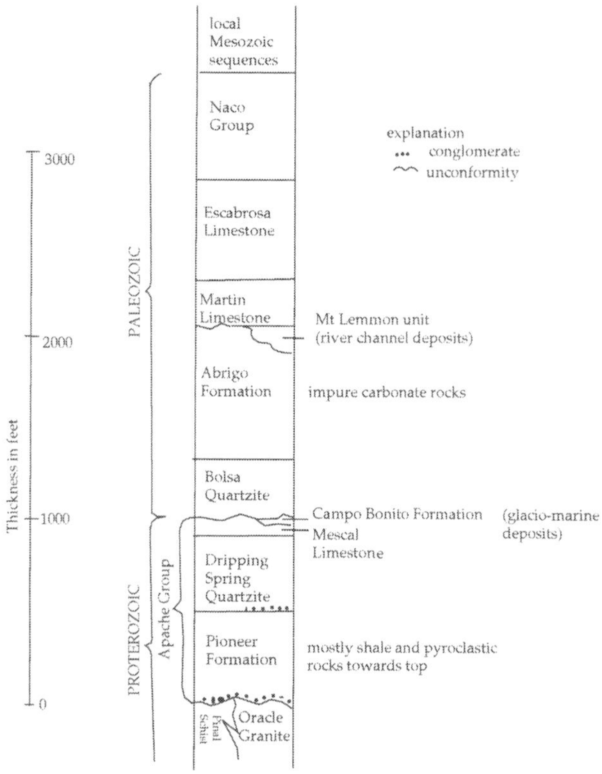

Figure 7 Simplified stratigraphic column for the study area.

PREVIOUS MAPPING

Approximately 70 percent of the study area had been mapped (fig. 3) with sufficient accuracy to form the basis of my own investigations. These maps are conceptually outdated to varying degrees, and of course none were compiled for the same purposes as those of this study, so all existing mapping was extensively field-checked and re-interpreted. New mapping is presented here for the remaining areas (pl. 1).

Published Mapping

Banks (1974) mapped the Mt. Lemmon (15') quadrangle at 1:62,500, i.e. Mt. Lemmon and Sabino Canyon 1:24,000 quadrangles of figure 3. This uncolored map is quite accurate, and represents admirable athletic ability. Depiction of structure is conservative; Banks' study initiated a period of great ferment in concepts of Catalina geology (see Banks, 1980, for his contributions), and his map reflects some but not all of the newer concepts.

Creasey (1967) mapped Mammoth (15') quadrangle at 1:48,000, i.e. Mammoth and Campo Bonito 1:24,000 quadrangles of figure 3. The upland part of Campo Bonito quadrangle is the main basis for plate 1 in

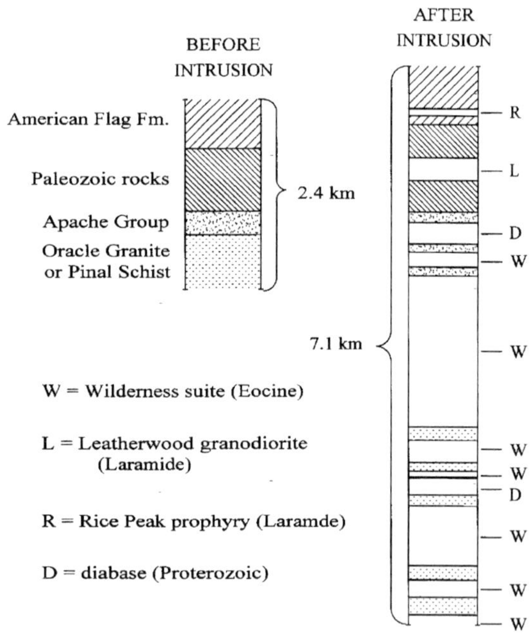

Figure 8 Total effects of Proterozoic through Eocene intrusions in the study area, shown in composite form. Stratigraphic levels of intrusions are diagrammatic but thickness is to scale.

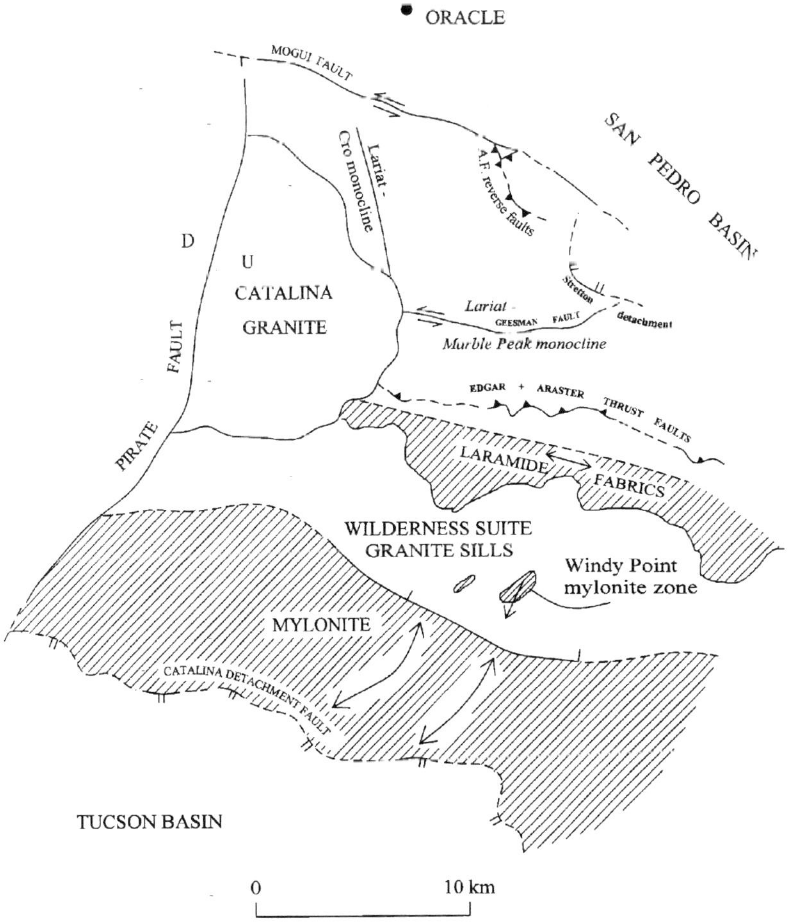

Figure 9 Summary map showing some major features of the Catalina range (i.e., the southern part of study area). Arrows show direction and plunge of lineation.

this area. An aeromagnetic map of the same area (but not at the same scale) is by Dempsey and Hill (1963).

The Mammoth 1:24,000 quadrangle was presented in greater detail by Creasey (1965), in a study emphasizing the San Manuel and Mammoth mining districts (see also Schwartz, 1953; Heindl, 1963; Thomas, 1966). These studies predate that of Lowell (1968) in the same area, which introduced new concepts of extensional faulting. Force and Cox (1992) presented a revised 1:24,000 map of part of the Mammoth quadrangle. Their map reflects contributions by Lowell (1968), Dickinson (1987, 1991), Dickinson and Shafiqulah (1989), and Weibel (1981), as well as their own work.

Creasey and Theodore (1975) is an open-file map of Bellota Ranch (15') quadrangle (scale apparently 1:30,500), i.e. Mt. Bigelow and Agua Caliente 1:24,000 quadrangles of figure 3. In my study area, their mapping was of reconnaissance nature in all but the northernmost portion, due to difficult access.

Dickinson (1993) is an open-file map at 1:24,000 of the San Manuel area, including parts of Mammoth, Campo Bonito, and North of Oracle quadrangles. This map emphasizes Tertiary units and structures.

Thesis and Dissertation Maps

Past mapping by students of the Geoscience Department of the University of Arizona contributed considerably to this study. Two of these maps (Suemnicht, 1977; Janecke, 1986) are also available from the Arizona Geological Survey. Only those maps plotted on topographic bases are discussed here, in approximate order of importance to this study.

Waag (1968) mapped the Mt. Lemmon and Mt. Bigelow areas at 1:6000. His map includes much of

Figure 10 Catalina forerange from the west, showing flatirons in south limb of forerange anticline, formed by a combination of leucocratic sills and parallel mylonitic foliation.

the high-elevation part of the study area. His emphasis was on detailed structural geology of Laramide deformation.

Peterson (1968) mapped the eastern Catalina forerange at 1:15,800. His units have been re-interpreted in this study, but his map pattern is an accurate guide to the distribution of some pre-mylonite lithologies, and to some structures in mylonitic rock.

Janecke (1986) mapped the Geesaman Wash area at 1:12,000. Her map supercedes Creasey and Theodore (1975) in that area. Some of her rock assignments have been revised, but her map patterns are accurate and she presented many hypotheses for testing.

Suemnicht (1977) mapped the headwaters area of the Canada del Oro, which drains much of the northern part of the Catalinas. He provided much information in an area of difficult access. Young (1988) mapped the pegmatitic Lemmon Rock unit at 1:6000. Pilkington (1962) made a reconnaissance geologic map at 1:15,000 of the east flank of the Catalina crest, a region of difficult access. Braun (1969) made a geologic map of the Marble Peak area, but did not differentiate the Paleozoic carbonate units. Hansen (1983) prepared a map of the western Black Hills. Wallace (1954) mapped an area just south of the Mogul fault. Some excellent thesis maps that were superseded by subsequent published maps include those of Bromfield (1950) and Ludden (1950) in the Peppersauce-Campo Bonito area, Pashley (1966) on the northern margin of the Tucson basin, and Weibel (1981) in the Black Hills. The work of Peirce (1958) has been superseded but in some ways never matched in the difficult area on the north side of the main range.

CONCEPTUAL STUDIES

Prior to studies published in Memoir 153 of the Geological Society of America in 1980, the age and nature of the crystalline rocks in the Catalinas were poorly known, due in part to a mylonitic overprint that was misunderstood. Older unpublished works that foreshadowed modern interpretations include that of C.F. Tolman and B.N. Moore in the first half of the century.

In Memoir 153, Keith and others (1980) described and proposed the currently accepted ages for the plutonic rocks of the area, based in part on earlier work by Shakel (1974, 1978). They first showed the Wilderness suite leucogranites as sub-parallel sills in older host rocks in their cross sections, and showed mylonitization as an overprint. In this report, much informal nomenclature dealing with plutons follows Keith and others. Elsewhere in the Memoir, Davis (1980) described the style of mylonitic deformation, and Banks (1980) described the composition and deformation-induced recrystallization of plutonic rocks.

Subsequently Anderson and others (1988), Davy and others (1989), and Palais and Peacock (1990) have discussed the age and emplacement depth of plutonic rocks in the Catalinas, and Naruk (1987), Naruk and Bykerk-Kauffman (1990), Reynolds and Lister (1990), and Guerin and others (1990) have debated the geometry of mylonitic deformation. Ruf and others (1988) studied magnetic properties of Catalina mylonites. Holt and others (1986) and Myers and Beck (1994) conducted gravity and seismic investigations of Catalina crust. Bykerk-Kauffman's (1990) thesis discussed Laramide deformation northeast of the range crest. Other topical theses pertaining to the eastern Catalinas are on the Mogul fault and Oracle Granite by Banerjee (1957), on Molino basin by Laughlin (1960), on contact metamorphism by Wood (1963), on Leatherwood "quartz diorite" by Hanson (1966), on pegmatites by Matter (1969), and on the petrography of the forerange by Sherwonit (1978).

Dickinson (1991) described the evolution of the mid-Tertiary tilted sequences that form the ends of the study area. In a sense this is a companion volume to his.

The larger mineral districts of the study area have been described. In the San Manuel area, Lowell's landmark paper of 1968 introduced the concept of low-angle extensional faults to the region, and Lowell and Guilbert (1970) established a geometry for alteration zones around porphyry copper deposits based on San Manuel. Force and others (1995) discuss the magnitude and history of tectonic rotation near San Manuel. Earlier descriptions of this district were by Schwartz (1953), Pelletier and Creasey (1965), and Thomas (1966). The Mammoth district was described by Peterson (1938), Creasey (1950), and Force and Cox (1992). Force and Dickinson (1994) describe some localities in these two districts. Peterson and Creasey (1943) and Braun (1969) described the Marble Peak district.

THIS STUDY

Field work for this study was conducted intermittently from 1989 through 1992; approximately 160 days were spent in the field. Except in a few circumstances, the work consisted of traverses with side excursions, i.e. contacts were generally not walked. Thus the map is still of a reconnaissance nature and a better job could be done. Geophysical coverage was insufficient to be of much help.

The geologic map (pl. 1; 1:48,000), though quite conventional in most respects, needs explanation in a few others. Where a protolith has been recrystallized and/or deformed, the protolith is nevertheless the mapped unit. The extent of mylonitization and some other varietal information is shown in plate 1B. Where an igneous unit has intruded another rock unit, the intrusive is the mapped unit where it comprises over 50 percent of the area, unlike the procedure locally followed by Banks (1974).

To the extent possible, structures of different ages are coded separately on the maps. Concealed contacts are extended into plutons as well as under cover. The cross sections are oriented NNE-SSW, sub-parallel to the direction of mid-Tertiary tectonic transport, to aid tectonic reconstruction. Plate 2 (1:12,000) shows the geology of some areas in greater detail.

No stratigraphic sections were measured in this study. Thicknesses are taken from the maps and cross sections except where measured thicknesses from previous studies are quoted.

Acknowledgments

My greatest debt in this study is to my wife Jane, who served as field assistant for the great majority of traverses, and drafted many of the figures. Others who assisted in the field were Leslie Cox, Mark Gettings, Prem Luthra, and Steve Pitts. Paul Damon, Steve Reynolds, and Bill Dickinson acquainted me with various parts of the area. Mark Barton, Larry Meinert, and many other field trip participants made valuable suggestions.

The U. S. Geological Survey supported this work as part of the Sierrita-Mogollon transect project. The Arizona Geological Survey and U. S. Bureau of Mines helped with compilation of locations for mines and prospects. My allies in geochronology, paleomagnetics and fluid inclusions were George Gehrels, Dan Unruh, Suzanne Baldwin, John Hagstrum, and Robert Kamilli; two of these investigations are reported in this volume.

S. C. Creasey provided a copy of his unpublished map that approximately covers the area of Oracle 1:24,000 quadrangle. Ann Bykerk-Kauffman contributed some field data in the area between the range crest to the Edgar fault. Paul Olsen and Kathryn Gregory contributed map information for plate 2A. Kevin Horstman helped with satellite images. Particular thanks go to James Etshoken and his U. S. Forest Service crew for trail revivification.

The manuscript was greatly improved by suggestions from K. A. Howard, S. R. Titley, B. B. Houser, Mark Barton and Jon Spencer. Map digitization and preparation was by Xaio Ling Qi, Averill Cate, Kris Ann Pizarro, and Robert Chaney, with supervisors Craig Wissler and Susan Tingley. Without Mark Barton, Joanne Newcomb, Susan Tingley, Bob Kamilli, and Meg Watt, this manuscript might still be in my filing cabinet.

DESCRIPTION OF ROCK UNITS

Rock units are described in this section from oldest to youngest in conventional manner. In the structure section, the effect of events are progressively removed from younger to older.

OLDER PRECAMBRIAN (EARLY TO MIDDLE PROTEROZOIC) ROCKS

Pinal Schist

Occurrence and description. The Pinal Schist, the oldest unit, is a far more abundant component of "basement" than previously recognized in the study area. It consists mostly of argillite to schist and meta-rhyolite, and commonly shows refolded fabrics. It previously has been confused with Abrigo Formation (mostly in Alder Canyon) and with phyllitic Pioneer Formation. However, Pinal can be distinguished in most cases by stratigraphic position and in others by its metamorphic segregations of quartz bands or the presence of two foliations. Proper identification of Pinal Schist is important because its distribution in the basement terrane unravels several structural problems in the study area (pl. 1B).

Pinal Schist forms the basement rock in a considerable area in the Canada del Oro drainage (Oracle quadrangle), where it is intruded by Oracle Granite and bounded on the south by younger intrusives, on the north by the Mogul fault, and on the east by unconformably overlying Pioneer Formation. In this area, the Pinal consists of meta-rhyolite and purple, green, and white phyllitic rocks with kinked foliation dipping steeply westward. Pinal also includes alternations of thin sandy and argillitic beds, probably of turbidite origin. An older foliation is apparently a Precambrian feature as it is cut by diabase dikes and unconformably overlain by the Apache Group. However, folding or kinking of foliation and overprinted lineations postdate the diabase and seem to be related to Laramide events. Quartz veins are common; some have comb-structured feldspar.

A window through a splay of the Geesaman fault (pl. 1) exposes Pinal Schist in Alder Canyon. Here the Pinal is mostly pelitic schist with quartz segregations (fig. 11, loc. 1). Immediately north of the Edgar thrust, Pinal Schist underlies Apache Group in several places (Bykerk-Kauffman, 1990). Pinal forms a small exposure in the bed of the south fork of Edgar Canyon (pl. 1, loc. 2). Judging from its abundance in float, Pinal Schist is exposed somewhere in the middle (main) fork of Edgar Canyon also. On the flanks of Kellogg Mountain, Pinal Schist underlies the basal Scanlan Conglomerate Member of the Pioneer Formation, and is intruded by minor Oracle Granite (fig. 12, loc. 3). The Pinal, which here consists of segregated quartz schist, is structurally bounded below by the younger Wilderness granite sill.

Together, these occurrences imply a continuous north-south belt of Pinal basement south of the Geesaman fault (pl. 1B) that may correlate with that in the Canada del Oro area north of the fault.

Correlation —The Pinal Schist is the oldest rock of the southeastern Arizona area. It is named for the Pinal Mountains south of the town of Globe, 70 km (45 mi) north of the study area. Deformation of the Pinal Schist clearly predates intrusion of Oracle Granite, and is thought to be at about 1650 Ma in the Pinal Mountains (Karlstrom and Bowring, 1988) and in the Little Dragoon Mountains to the southwest (Silver, 1978). The subequal abundance of rhyolite and clastics in the Pinal of the study area is anomalous for the immediate region.

Figure 11 Pinal Schist in Alder Canyon, showing folded quartz segregations, at locality 1.

Oracle Granite and Related Rocks

Description. Near the town of Oracle, for which the granite was named by Peterson (1938), Oracle Granite consists of a coarse porphyritic rock (fig. 13) with biotite the principal mafic mineral. The phenocrysts are of perthitic potassium feldspar, commonly 2 cm by 3 cm in cross section, with characteristic plagioclase rims as in rapakivi texture. Plagioclase, quartz, and fine potassium feldspar in 2 to 3 mm grains form most of the "groundmass". Biotite, partially to completely chloritized, averages about 10 to 13 percent (Creasey, 1967). Plagioclase is commonly sericitized. Magnetite at nearly one percent makes the rock moderately magnetic. Other accessory phases are apatite, locally sphene, and zircon (table 1).

In the Oracle area, bounded by the Mogul fault on the south and Tertiary sedimentary rocks on the north and east (pl. 1), Oracle Granite as described above forms about 90 percent of bedrock outcrop. Several variations that are genetically related to Oracle Granite also occur in this area and will be described here.

Figure 12 Pinal Schist and unconformably overlying Scanlan Conglomerate Member of Pioneer Formation with stretched pebbles, on northern flank of Kellogg Mountain, Mt. Bigelow quadrangle (loc. 3).

Leucocratic granite facies. A more equigranular, medium-grained, biotite-poor facies of the Oracle Granite forms diffuse bodies too small to be mapped throughout the Oracle area, but occupies most of the pediment outcrops between Peppersauce Wash and the Mogul fault in the San Pedro basin. Creasey (1967), who was unaware of these outcrops, called this rock type alaskite elsewhere and related it to aplite, citing gradational contacts and similarity in orientation.

Aplite and pegmatite. Closely associated bodies of aplite and pegmatite form tabular bodies up to 10 m (35 ft) thick trending NNW throughout the type region. They are common also in the San Manuel district on both sides of the San Manuel fault. The aplites are saccharoidal and massive. Pegmatites occupy the same tabular bodies in close association with aplite. Commonly these pegmatites consist of graphic quartz-feldspar intergrowths, separate blue-gray quartz, and muscovite (mostly deformed). They commonly contain tourmaline of the schorl variety as masses up to 10 cm.

Aplite cuts neither Proterozoic diabase nor Laramide porphyry as shown by Creasey (1965) in Mammoth quadrangle. Both these rocks are younger than aplite. In the Oracle area are a few irregular dikes of magnetic blue-quartz porphyry that seem likely to be Precambrian, but they cut aplite.

Granodiorite. At the northern end of the study area, Creasey (1965, 1967) differentiated a darker, more equigranular granodiorite from the Oracle Granite. Creasey thought the granodiorite was the older.

Occurrences between the Mogul fault and the range crest. Oracle Granite and Pinal Schist form the basement for the overlying supracrustal sequence of later Precambrian through Tertiary age, mostly dipping northeast, between the Mogul fault and the range crest (pl. 1B). The largest exposure of Oracle Granite in this interval forms the lowland in the core of the southeast-trending Lariat anticline just north of the Geesaman fault (fig. 14, pl. 1). Intrusive contacts of Oracle Granite in Pinal Schist trend northeast but are poorly exposed.

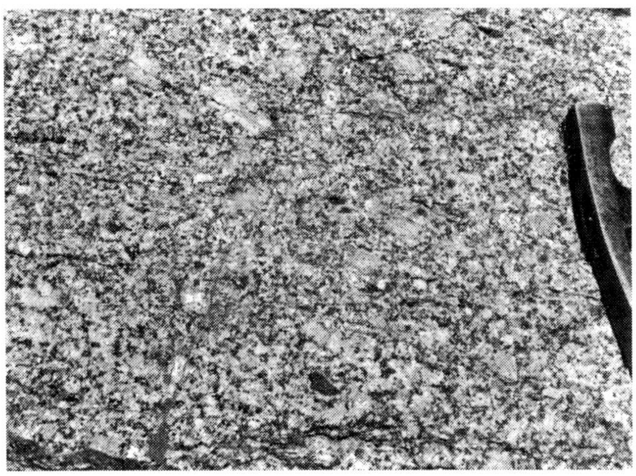

Figure 13 Porphyritic texture of Oracle Granite from the type region.

The Oracle Granite is closely comparable to that of the Oracle area in this region, though locally it may show slight foliation or protoclastic texture, due to the deformation that has affected all the pre-Tertiary rocks in this block. Aplite and pegmatite are somewhat less common in this area than north of the Mogul fault.

Occurrences among younger granites south of the range crest. The leucocratic Wilderness granite sill and related pegmatites of Eocene age have intruded along

Table 1 Chemical and modal analyses of fresh and mylonitic Oracle Granite.

A. Chemistry
(n.d., not determined; *, all iron reported as one oxide; +, analyzed by ICP; remainder by atomic absorbtion)

	Type area	Slightly deformed		Mylonitic			
	1+	**2**	**3+**	**4+**	**5+**	**6+**	**7+**
SiO_2	n.d.	70.2	n.d.	n.d.	n.d.	n.d.	n.d.
Al_2O_3	11.5	12.0	13.4	14.6	13.1	13.1	12.8
Fe_2O_3	5.4	0.72	7.1	9.2	6.2	4.8	5.6
FeO	*	1.5	*	*	*	*	*
MgO	0.8	0.82	1.7	2.2	1.2	1.0	1.2
CaO	2.2	0.85	1.0	3.1	2.2	2.0	2.5
Na_2O	2.1	2.17	2.1	1.4	2.1	2.6	2.5
K_2O	2.5	4.55	2.8	1.9	2.8	3.1	2.5
P_2O_5	0.5	0.17	0.1	0.6	0.3	0.2	0.2
TiO_2	0.6	n.d.	0.9	1.4	0.9	0.7	0.9

B. Modes
(n.d. not determined; tr, trace)

	Type area		Mylonitic	
	8	**9**	**10**	**11**
Quartz	26	33	32.4	31
K-feldspar	21	29	10.6	6
Plagioclase	38	28	33.8	31
An value	30+5	22-34	20	20-30
Biotite	n.d.	9	21.4	29
Apatite	n.d.	n.d.	0.6	tr
Opaque	n.d.	n.d.	1.0	1

Descriptions
1. Sample 155844 of Coney and Reynolds (1980), fresh, from type locality.
2. As above, sample 155843, weakly deformed, from upper Bear Canyon.
3. As above, sample 155841, from upper Bear Canyon.
4. As above, sample 155835, from Molino basin.
5. As above, sample 155819, from Soldier Canyon.
6. As above, sample 155915, from near Babat Duag overlook.
7. As above, sample 155916, from Molino Canyon.
8. Sample from near Oracle, reported by Creasey (1965).
9. Average of 38 samples from near Oracle, reported by Banerjee (1957).
10. Sample ML62D, mylonitic, from Catalina forerange, reported by Banks (1980).
11. Average of 24 samples, mylonitic, from Catalina forerange, reported by Sherwonit (1974).

and thereby split open the unconformity between the Oracle Granite and the Pioneer Formation in the main range, and structurally lower leucogranites of the Wilderness suite are emplaced within Oracle (pl. 1). Oracle forms numerous screens and xenoliths in this region. One map unit (Twi of plate 1) is defined by the abundance of xenoliths, mostly of the Oracle. Plate 1B shows the distribution of Oracle xenoliths in younger granites. Oracle xenoliths, some enormous, are common near the top of the Wilderness granite, including parts of its upper Lemmon Rock pegmatitic facies. Such xenoliths were the basis of the map unit "(Or)" of Banks (1974) below Summerhaven; plate 1 shows this area as Lemmon Rock pegmatite-aplite. Xenoliths of deformed Oracle in little-deformed Wilderness granite form the eastern face of Green Mountain (fig. 15, loc. 4). In only two areas was Oracle Granite found structurally above the Wilderness granite sill. One is the flank of Kellogg Mountain (loc. 3), described above, where it occurs with Pinal Schist below the Pioneer Formation. The other is along Sabino Canyon near Marshall Gulch (best exposed along Sunset Trail, fig. 16A, loc. 5).

Oracle Granite is still easily recognizable in this area, as deformation has been insufficient to destroy its characteristic texture except on Green Mountain. Were it not for the younger Wilderness granite sill, this area would show Oracle in the characteristic basement position it occupies elsewhere in the study area and throughout the region.

Mylonitic facies. In the Catalina forerange and immediately adjacent parts of the main range, Oracle Granite forms thick screens between Wilderness suite leucogranite sills, but here all the rocks have been mylonitized, and the Oracle preferentially so. In this area the Oracle consists of white feldspar in a dark mylonitic paste. This is the darker rock exposed along the Catalina Highway from its base into Molino Canyon, interlayered with leucogranite and pegmatite of the younger Wilderness suite. Some geologists have doubted that this rock should be called Oracle Granite, because it is so much darker than in its type locality. The following points are made to reinforce the correlation:

1) The mylonitic dark rock passes upward structurally into the screens of the main range that are easily recognized as Oracle. Other structural blocks containing recognizable Oracle then complete the spatial transition northward to the type region. Reconstruction of the pre-Wilderness position of the dark mylonite results in the correct stratigraphic-structural position of Oracle Granite.

2) The porphyroclasts of white feldspar are the same size and perthitic composition as the phenocrysts in Oracle Granite, and commonly show relict plagioclase (rapakivi) rims.

3) Radiometric ages on zircon show that the two rocks are about the same age (Shakel and others, 1977; Reynolds and others, 1986).

4) Chemical analyses (in Reynolds and others, 1980) show that the chemical compositions, including the iron and magnesium contents, of mylonitic and typical Oracle are about the same (table 1). Modal analyses (table 1) show that most mineral constituents are about the same except for biotite and potassium feldspar; biotite is higher in mylonite, apparently at the expense of fine-grained potassium feldspar in fresh

Figure 14 Oblique view of the Lariat anticline east of Oracle Ridge. Annotations as for plate 1; Gf is Geesaman fault.

Figure 15 Xenoliths of deformed Oracle Granite in little-deformed Wilderness granite on the eastern face of Green Mountain.

undeformed rock. Mylonitization has also smeared out biotite to a mylonitic paste that gives the rock a dark color.

Age and correlation. Radiometric ages of primary Oracle Granite range from 1351 to 1430 Ma, based on a variety of methods (Reynolds and others, 1986). The term Oracle Granite is used throughout the Rincon-Catalina-Tortolita area, and immediately to the north. Oracle Granite compares closely with other granites of the same age such as the Ruin Granite of the Globe area. Other nearby areas have Precambrian granitic rocks in the same general age range as Oracle Granite, such as the Continental Granodiorite of the Sierrita Mountains.

Figure 16 Unconformity between underlying Oracle Granite and overlying Scanlan Member, A. On the west flank of Oracle Ridge (loc. 6), B. On the east flank of Sabino Canyon below Summerhaven, in a screen engulfed in Lemmon Rock pegmatite (loc. 5). A dike of porphyritic Leatherwood is intruded along the unconformity. The unconformity is little-deformed in A, badly deformed in B.

Summary of Older Precambrian Rocks

The basement on which supracrustal sequences of later Precambrian and Paleozoic rocks of the study area were deposited consists mostly of Oracle Granite in the study area. Occurrences of Pinal Schist form one or possibly two main belts, separated by the Geesaman fault and truncated by the Mogul fault (pl. 1B).

The leucocratic facies of Oracle Granite on the pedimented margin of the San Pedro basin may represent a sizable separate pluton, and the NNW-trending aplites and pegmatites may be related more directly to this pluton than to normal Oracle Granite. A possibly correlative leucocratic facies of Oracle south of the Mogul fault is described by Bykerk-Kauffman (1990, p. 50).

YOUNGER PRECAMBRIAN (MIDDLE PROTEROZOIC) ROCKS—APACHE GROUP AND DIABASE

The Apache Group in the study area consists from base to top of Pioneer Formation, Dripping Spring Quartzite, and Mescal Limestone (fig. 7). The total thickness of the group is as much as about 275 meters, but it apparently thins toward the southwest. This sequence and older rocks have been intruded by Precambrian diabase.

Pioneer Formation

The basal formation of the Apache Group is the Pioneer Formation (in the sense of Willden, 1964, and Creasey, 1967, including the basal Scanlan Conglomerate Member; fig. 16).

Distribution and description. The area where the Pioneer Formation is best represented as a sedimentary rock is between the Mogul and Edgar faults. Here it is only slightly recrystallized and deformed, even adjacent to voluminous intrusions of Rice Peak porphyry. Pioneer Formation is also widely distributed along the crest of the main range, but there it is metamorphosed and badly deformed, locally to tight recumbent folds, and its structural base is characteristically against younger Wilderness granite.

From base to top, the Pioneer Formation consists of 1) the basal Scanlan Conglomerate Member, 2) a sandstone that Creasey (1967) referred to as the middle member, and 3) a thick sequence of siltstone and pyroclastic rocks. The overall thickness implied by Creasey is 450 to 510 feet, or 135 to 155 m. A definitive thickness is difficult to obtain, as the Pioneer Formation is the preferred host of diabase sills.

Basal unconformity and Scanlan Conglomerate Member. The Pioneer Formation rests unconformably on older Precambrian rocks described above, with Scanlan Conglomerate Member as its basal unit. Where best exposed and least deformed, in the block between the Mogul and Geesaman faults, the unconformity can have either of two characters: it can be sharp (figs. 12, 16) or may consist of a zone as wide as 10 m in which the Oracle Granite loses its feldspar but not its texture, overlain by pebbly coarse sandstone that is poorly sorted at its base. This transition probably represents a paleosol, but slight deformation and recrystallization discourage detailed study. On Sunset Trail (fig. 16A, loc. 5), where appreciable metamorphism has occurred, a probable regolith on Oracle has recrystallized to sericitic schist.

In some sections, the basal lithotype of the Scanlan is a gritty sand or a bouldery silty sand, and rounded pebbles appear in an overlying horizon (fig. 17, loc. 4). Near Bat Well on Geesaman Wash (loc. 8), a granular phyllite up to 3 m thick separates Oracle Granite from Scanlan. These basal beds are probably transported soils.

The Scanlan Member in my experience ranges in thickness from about 10 cm to about 10 m. Given this range in thickness, however, it is remarkably persistent — in no intact section was it found to be absent, over many kilometers of outcrop length and well distributed areally due to structural repetition. Even along the crest of the main range, where the Wilderness granite sill has been intruded along the unconformity, Scanlan commonly rests structurally on Wilderness granite.

The Scanlan typically consists of white well-rounded pebbles, mostly of quartz, in a sand matrix that is most commonly light gray (fig. 16). This character seems to vary little whether the underlying unconformity is on Oracle Granite or Pinal Schist (figs. 16, 17), though one locality where angular pebbles include Pinal Schist occurs in the Canada del Oro area. On the north flank of Kellogg Mountain in the main range, rounded pebbles in the Scanlan have been stretched to great length (fig. 12, loc. 3), and this phenomenon has been noted to lesser degrees at a number of other localities. Where the Scanlan is metamorphosed, clast-matrix boundaries may be indistinct (fig. 16b, loc. 5).

Intermediate sandstone. A sandstone about 30 m thick overlies but grades into Scanlan Conglomerate Member (Creasey, 1967) in part of the study area. The sandstone is commonly cross-bedded and locally pebbly, but the conglomerate bed in sandstone described

Figure 17 Unconformity of Pioneer Formation on Pinal Schist (Xp) in the bed of Canada del Oro, Oracle quadrangle (loc. 7). Gritty sand (Yp) underlies Scanlan Conglomerate.

Figure 18 Sandy cross-bedded Pioneer Formation middle member in Atchley Canyon, apparently derived from the east (loc. 9).

by Creasey on Oracle Ridge is actually a structural repetition of the Scanlan. In some sections toward the eastern end of the Lariat anticline, the sandstone is absent and fine-grained upper Pioneer rests directly on Scanlan. However, south of the Geesaman fault, the sandstone member is locally thick (fig. 18); the distribution of the sandstone member may be explained by the fault itself.

Figure 19 Typical laminated siltstone of the upper part of the Pioneer Formation, from the canyon below Mary West mine (loc. 10).

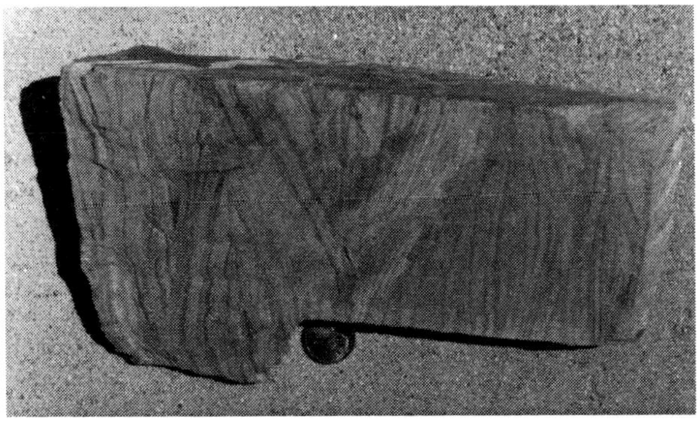

Figure 20 Specimen of silty slate in upper Pioneer Formation from Juanito Spring (loc. 11). Photo perpendicular to slaty cleavage; specimen is 1.5 cm thick.

Upper member. Most of the thickness of the Pioneer Formation consists of fine-grained rocks of the upper member. Indeed the formation was formerly called Pioneer Shale when Scanlan Conglomerate Member was not included. Creasey (1967) measured the thickness of the upper fine-grained member as 365 feet (110 m). Natural exposure is generally rather poor; however, where this part of the formation is free of diabase sills, as along Oracle Ridge south of Rice Peak, it can be a resistant lithology.

In the block between the Mogul and Geesaman faults, where its original character is best preserved, this fine-grained upper member of the Pioneer consists mostly of finely-bedded siltstones (fig. 19), which may be either maroon or gray-green in color, and lesser feldspathic fine sandstone. Graded bedding is common and cross-lamination (fig. 20, loc. 11) less common. Locally, bedding is obscured by cleavage. True slate (fig. 20) occurs in some localities.

The upper part of the Pioneer is the preferred host of younger diabase sills throughout the study area, probably because of its mechanical weakness compared with other Precambrian rocks. Diabase intrusions contribute to poor exposure of the upper part of the Pioneer, because the diabase is not generally resistant to erosion. In the crestal area of the range, diabase and phyllite were deformed and metamorphosed together so intricately that in some outcrops, digitate boundaries between rock types can be defined only by the texture. Metasomatism seems likely. Pioneer plus diabase was generally mapped as a single unit (pl. 1), partly because of this problem.

Pyroclastic rocks. Creasey (1967) noted that feldspars in the upper member are crystal fragments, presumably of pyroclastic origin, and that the siltstones contain vitric fragments. I find that in certain areas, lapilli, vitric, and ash-flow tuffs form entire stratigraphic intervals varying from about 10 m in the middle of the upper fine-grained member of the Pioneer, to sections more than 100 m thick comprising the entire upper member of the Pioneer and resting directly on Scanlan. The pyroclastic component of the formation seems to reach a maximum in both grain size and thickness in an area just north of the Geesaman fault in the Lariat anticline area (pl. 1B), suggesting proximity to a Precambrian volcanic center. In the eastern part of this area, the pyroclastic rocks were previously mapped as Rice Peak porphyry, but they are clearly cut by Precambrian diabase (fig. 21, loc. 12). South of the Geesaman fault, Pioneer pyroclastics are locally coarse or thick (loc. 13), but their extent is poorly defined (pl. 1B).

Ash-flow tuffs are the most abundant type of pyroclastic rock in the Pioneer. Typically they contain rolled beta-quartz-eye phenocrysts and flattened pumice fragments in an altered feldspathic groundmass (figs. 22, 23). These rocks have variable SiO_2 contents, and are so potassic that metasomatism is likely (table 2). The composition of fine clastic lithotypes is similar to that of the pyroclastics except for higher Fe_2O_3/FeO ratios. This suggests that either the fine clastics are reworked pyroclastics, as Creasey suggested, or that metasomatism has overwhelmed both lithotypes .

Phyllitic areas. Along the crest of the main range, the Pioneer Formation has been severely deformed and metamorphosed, and its upper part has been transformed into light to dark gray, micaceous phyllite with porphyroblasts of magnetite, garnet, and other minerals. Moderately high pyrite content as porphyroblasts has made many outcrops rusty. Cleavage cuts bedding, and in a few outcrops, two foliations are present, as at Inspiration Rock (loc. 17).

Origin and correlation. The areal persistence of the Scanlan Conglomerate Member suggests that it is a transgressive lag deposit rather than a channel deposit in the study area. This is consistent with the thick fine-grained clastic facies of the upper part of the Pioneer, which seems likely to be marine. Pyroclastic volcanism of silicic composition during accumulation of the upper member was apparently centered on the present Lariat anticline.

The Pioneer Formation is best represented in the Globe-Miami area; north and east of there it pinches out against older Precambrian rocks (Wrucke, 1989). In no other area to date has a thick section of pyroclastic rocks been reported within the Pioneer.

Dripping Spring Quartzite

Description. The distribution of Dripping Spring Quartzite in the study area is similar to that of the Pioneer Formation, which it overlies. Being more resistant to both deformation and recrystallization, the Dripping Spring shows less contrast across the study area in these respects than does the Pioneer.

The Dripping Spring Quartzite consists of the basal Barnes Conglomerate Member (included by Granger and Raup, 1964), and an upper interval of fine quartzite. The total thickness of the formation is about 100 m. It rests with apparent conformity on Pioneer Formation in the study area.

Barnes Conglomerate Member. Because of its marked resistance to erosion and its restricted thickness (0 to 20 m), the Barnes Conglomerate Member where

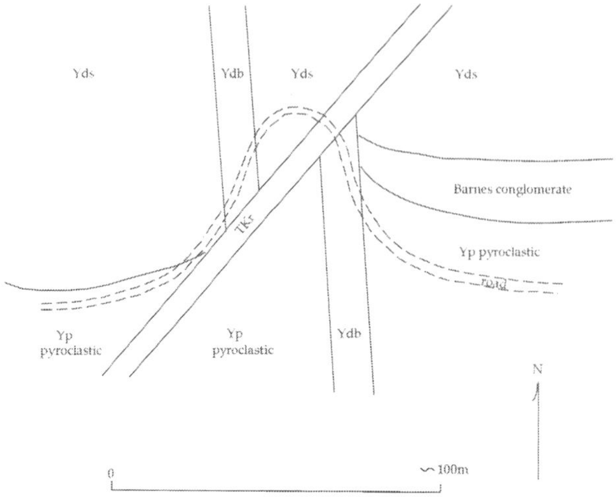

Figure 21 Diagrammatic map of geologic relations among lenticular Barnes Conglomerate Member of Dripping Spring Quartzite (Yds), largely pyroclastic Pioneer Formation (Yp), Precambrian diabase (Ydb), and Rice Peak porphyry (TKr) at locality 12.

Figure 22 Welded ash-flow tuff with quartz eyes in the upper part of the Pioneer Formation, from Corregedor mine area (table 2, loc. 14). Scale in cm.

Figure 23 Photomicrograph of ash-flow tuff from the upper Pioneer Formation, showing pumice fragment to right. Field of view 6 mm, plane light.

Figure 24 Sandstone layer at base of Barnes Member, from canyon below Mary West mine (loc. 10). Bottom of lens cap at unconformity on Pioneer Formation siltstone.

Figure 25 Cross-bedding in Dripping Spring Quartzite, from Atchley Canyon.

Figure 26 Mudcracks in argillite beds of upper Dripping Spring Quartzite, from the canyon below Mary West mine.

present is a useful marker bed on the northern flank of the range. It is not as widely distributed as the Scanlan, however; Barnes is absent in about half the sections I observed.

Conglomerate of the Barnes Member consists of well-rounded pebbles or cobbles of various siliceous types—quartz, fine-grained gray quartzite, jasper—in a sandy maroon matrix (loc. 18). Where the Barnes is metamorphosed, the matrix is gray. As with the Scanlan, stretched pebbles are common. Waag (1968) and Bykerk-Kauffman (1990) have recorded localities where the pebbles have been stretched beyond easy recognition.

The base of the Barnes is an erosional surface cut into the fine-grained upper part of the Pioneer. In some localities, the basal bed is a gritty sand (fig. 24, loc. 10).

Table 2 Chemical composition of upper member of Pioneer Formation, including pyroclastic facies. All analyses by WD XRF, by D. F. Siems and T. R. Peacock.

	Pyroclastic facies		Fine clastic facies	
	1	**2**	**3**	**4**
SiO_2	57.4	70.5	69.9	70.3
$Al2O_3$	16.9	15.6	15.4	15.3
Fe_2O_3	3.87	2.45	3.52	4.89
FeO	1.22	0.83	0.43	0.32
MgO	2.93	0.74	1.16	0.50
CaO	7.81	0.27	0.13	0.05
Na_2O	0.62	<0.15	<0.15	<0.15
K_2O	6.08	5.25	5.93	5.12
P_2O_5	0.17	0.20	0.1	0.08
TiO_2	0.67	0.57	0.44	0.48

Descriptions
1. Sample FT EJ (loc. 14), quartz - K-spar ignimbrite from Corregedor mine area.
2. Sample FT EE (loc. 15), quartz ignimbrite, Control road
3. Sample FT 772B (loc. 9) laminated siltstone with pseudomatrix, below Mary West mine
4. Sample FT BP (loc. 16), siltstone, Oracle Ridge

Quartzite. Barnes Conglomerate Member passes up gradationally into fine feldspathic quartzite that forms most of the formation. Creasey (1967) noted that the quartzite can be divided into two units. The basal unit, about 60 m thick, is thicker-bedded, cross-bedded (fig. 25), and slightly coarser grained (medium to fine sand). The basal sand may be coarse where the Barnes is missing. The upper unit (about 30 m) is finer-grained and contains closely spaced slaty to micaceous partings and thin laminated beds that locally contain mud-cracks (fig. 26). The upper unit is rusty-weathering in some sections. In Atchley and Edgar Canyons and Canada del Oro, the uppermost Dripping Spring consists of a thick interval of pyritic dark laminated argillite (locs. 13, 19, 20), probably once continuous across the Geesaman fault (pl. 1B). Waag (1968) was able to map discrete phyllite units toward the top of the formation. In several areas, the Dripping Spring is intruded by diabase sills, mostly in the upper unit.

Origin and correlation. The irregular distribution of the Barnes Member suggests lenticularity and a fluvial-channel origin. Its gradual transition upward into cross-bedded quartzite and mud-cracked argillite suggests a fluvial origin for most of the Dripping Spring Quartzite also. The dark pyritic laminated argillite facies was probably deposited in stratified alkaline lakes. The jasper among the clasts in Barnes may reflect the nearby volcanism of the underlying Pioneer.

The Dripping Spring Quartzite is widely distributed to the north in the Globe-Miami area and the Mogollon Rim (Wrucke, 1989). It is named for the Dripping Spring Mountains, which is about 50 mi (80 km) to the north.

Mescal Limestone

Distribution and description. Mescal Limestone, the uppermost unit of the Apache Group, is present in the study area along the crest of the main range near Mt. Bigelow, where it is deformed and recrystallized, between the Araster and Edgar thrust faults, and in the Canada del Oro drainage. Mescal is missing, probably because of Precambrian erosion, elsewhere in the study area. Its identification along the range crest is based on position between Dripping Spring and Bolsa Quartzites, intrusion by diabase, and dissimilarity to any lower Paleozoic limestones. The Mescal along the crest occurs in discrete areas due partly to thrust faults and partly to the vagaries of intersection of nearly parallel topography and deformed bedding.

The Mescal in the crestal area consists mostly of marble, laced with quartzite (after chert) both along and across bedding. Bedding in some intervals consists of fine alternations of marble and phyllite (fig. 27, loc. 21). Recrystallization has produced some calc-silicate minerals, but not nearly so much as in the younger Abrigo Formation. The total stratigraphic thickness is uncertain because of structural complexity, but its structural thickness is locally as great as 60 m.

In the Canada del Oro drainage, south of the Mogul fault, Mescal Limestone forms the core of the Oro syncline (pl. 1). The structure is complex, but recrystallization is modest except along the basal contact with Dripping Spring Quartzite, where sulfide minerals and epidote are abundant (loc. 22). Cherty limestone and finely banded limestone are the most common lithologies. The apparent thickness is as great as 60 m, but the upper contact is not preserved. Between the Araster and Edgar faults, the Mescal is cherty marble.

Origin and correlation. The Mescal was probably formed in a marine shelf environment. The name is from the Mescal Mountains in the Globe-Miami area to the north, where a far more extensive sequence including algal and hypersaline facies is preserved (Wrucke, 1989). The upper parts of the Mescal as reported in other areas seem to be missing in the study area. The original distribution of Mescal in the study area was probably more extensive; there is evidence of post-Apache erosion surfaces in some areas where Mescal is missing, as is discussed below with Campo Bonito Formation.

Diabase

Locally voluminous diabase is intrusive into the Apache Group and older Precambrian rocks, but unconformably overlain by Cambrian rocks. Where fresh, the diabase shows ophitic and subophitic textures of clinopyroxene, locally altered to hornblende, and andesine (fig. 28). Olivine is preserved in some areas, and may exceed 10 percent. Minor quartz and potassium feldspar occur in local granophyres. Magnetite averages about one volume percent, enough to render fresh diabase the most magnetic rock unit of appreciable size in the study area. Commonly magnetite is skeletal and locally altered to hematite and sphene. Grain size toward the center of the larger diabase bodies is as much as 2 to 3 mm (where it is technically a gabbro), or greater in local pegmatites. Major-element analyses of diabase from the study area are shown in table 3. In the southern part of the study area, the diabase is every-

Table 3 Chemical composition of Proterozoic diabase from the study area, and comparison with diabase from nearby ranges; *, all iron reported as one oxide.

	1	2	3
SiO_2	49.5	n.d.	47.7
Al_2O_3	14.1	16.3	16.8
Fe_2O_3	*	13.8	1.5
FeO	12.1	*	10.0
MgO	8.15	6.7	7.0
CaO	12.0	9.2	8.7
Na_2O	1.96	2.8	2.8
K_2O	0.83	1.1	1.0
P_2O_5	0.11	0.2	0.3
TiO_2	0.98	1.9	1.7

Descriptions

1. Sample FT 90 from Purcell window (Kalamazoo) area, San Manuel district (loc. 26). Analysis by XRF for this study supplied by Daniel M. Unruh.
2. Average of five ICP-AES analyses by P. H. Briggs, samples FT 224, 233, 241B, 766A, and BE from the northern part of the study area (locs. 23, 24, 27, 28).
3. Average of nine rapid-rock analyses, reported by Wrucke (1989) from Aravaipa and Salt River canyons and Sierra Ancha.

where altered to chlorite-, biotite-, and (or) amphibole-bearing assemblages. Three main habits of diabase occur:

Dikes in Pinal Schist. Fewer diabase dikes intrude Pinal Schist than other host rocks. Some dikes are oriented approximately N-S but are deformed with the foliation of the schist, presumably by Laramide deformation.

Figure 27 Fine bedding in impure marble formed from Mescal Limestone at Bear Wallow (loc. 21).

Figure 28 Photomicrograph of fresh diabase north of American Flag Hill (loc. 23). Field of view 6 mm, plane light.

"Dikes" in Oracle Granite. Diabase bodies (informally called dikes), mostly striking NNW with northeast dips to 30 to 50°, are found throughout the areas of Oracle Granite and related rocks. Deformed metadiabase can be recognized even in the mylonitic Oracle of the forerange. Individual dike thicknesses range from less than a meter to over 100 m; aggregate thicknesses average about 250 m. The largest single body is that which forms American Flag Hill (loc. 23); the true thickness of this body is probably less than 200 m. Dike margins are chilled, and grain size coarsens to gabbro in some dike interiors. Dike-margin contacts are commonly fractured, and the quartz + carbonate veins emplaced in these fractures are discussed under mineral resources. Howard (1991) showed that the diabase dikes were subhorizontal sheets in this region prior to mid-Tertiary tilting.

The diabase dikes cut aplite and pegmatite bodies and are cut by Laramide porphyries throughout the study area. Creasey (1967) wrote of older and younger generations of diabase, but he was clearly mistaken in the two locations in Mammoth quadrangle where he showed such relations in the study area. In the Kalamazoo area an aplite is shown cutting a diabase body that does not exist; this was probably a drafting error. The body near Red Hill that he shows as cutting Laramide porphyry is not a diabase. Thus in the study area there is field evidence of only one period of diabase intrusion.

Sills in Apache Group. Individual diabase sills up to 200 m thick intrude the Apache Group, especially in the upper part of the Pioneer Formation and secondarily in the upper part of the Dripping Spring Quartzite (loc. 24). Small diabase bodies may be found almost anywhere in the Apache Group. On plate 1, diabase in Apache Group has generally not been separated because it is ubiquitous. Diabase structurally expanded the Apache Group 300 m or more in many parts of the study area.

The diabase sills are least altered and deformed north of the Edgar fault. The sills are chilled on the margins and coarsen inward to gabbro in the thicker bodies. Locally the margins are fissile and chloritic, but most interiors of thick sills seem undeformed. Where the wallrocks are argillaceous, they locally show a few meters of spotted hornfels adjacent to the diabase contact.

Along and north of the crest of the main range, equally voluminous diabase at the same horizons of Apache Group has most commonly been converted to phyllite or schist composed mostly of biotite with lesser feldspar. Less deformed sill interiors show biotite and/or hornblende in domains separated by plagioclase that represent flattened sub-ophitic textures (loc. 25).

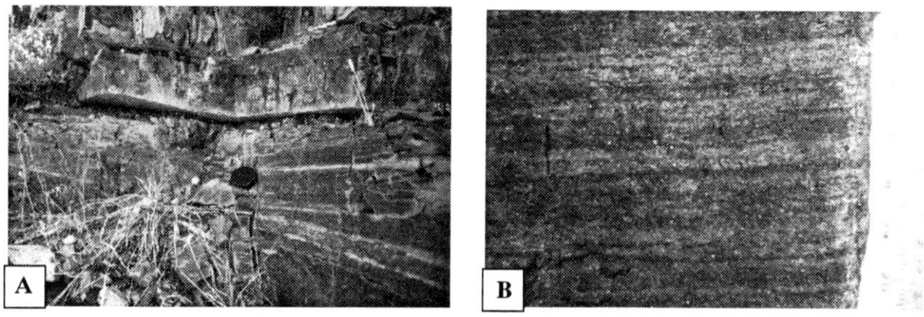

Figure 29 Laminated sand-silt interbeds in upper Campo Bonito Formation in Southern Belle mine area (loc. 29). A. Outcrop showing unconformity with overlying Bolsa Quartzite (above lens cap), B. Hand specimen (scale in cm).

Figure 30 Conglomerate near the base of the Campo Bonito Formation near Southern Belle mine (loc. 29).

Age and correlation. Diabase of 1,040 to 1,150 Ma age is widely distributed in south-central Arizona (Wrucke, 1989). It is approximately coextensive with the Apache Group that it intrudes, suggesting a rift-related origin for the distribution of both; however, Howard (1991) suggests a compressive regime for diabase emplacement. Compositions reported by Wrucke from other parts of the region are similar to those of the study area (table 3).

PROBABLE LATE PROTEROZOIC UNIT

Campo Bonito Formation (new)

Distribution and description. Creasey (1967) included in the Bolsa Quartzite of Cambrian age a unit of "maroon siltstone" that lies below the quartzite comprising the main part of the Bolsa. Stoyanow (1936) considered this unit Precambrian. The maroon siltstone, up to 30 m thick, is quite restricted in distribution, being present in an area about 3.5 km long and 2 km wide (pl. 1) south of the Mogul fault. In parts of the Southern Belle district within this area, the unit is missing, probably due to bedding-parallel faults. Outside this area, however, Paleozoic rocks lie directly on Apache Group. The restricted distribution of the maroon siltstone in the study area probably reflects the primary depositional distribution.

In two areas, the same stratigraphic interval occupied by the maroon siltstone is instead occupied by a diamictite unit 20 m thick. Available evidence suggests that these two units are related in age and origin.

Maroon siltstone. This unit appears massive through most of its thickness, probably due to homogeneous composition. Where thin sand layers are present the rock is seen to be finely laminated (fig. 29). These sand layers are most common where the unit is thin, and toward the top of the unit, which also shows lenticular bleached zones. A conglomerate with angular clasts (fig. 30) is present at the base of the unit interbedded with varved fine sand. Such conglomerate beds occur stratigraphically higher where the unit is thin. Diabase clasts are locally present (Creasey, 1967).

The unit apparently is finest grained and contains the most iron-oxide matrix in the canyon below Southern Belle mine (loc. 29). Relatively thick sections were seen in that canyon and Peppersauce canyon (loc. 30). A condensed section (about 8.5 m or 28 ft) can be seen on a ridge between Southern Belle mine and the Campo Bonito district (loc. 31).

Sedimentary structures other than lamination are rare in the maroon siltstone unit. Trace fossils were not seen. Body fossils are not reported by Stoyanow (1936) or Creasey (1967), nor did I find any.

The maroon siltstone contains poorly sorted angular to subrounded grains of quartz, plagioclase, mica, and K-feldspar in an iron-oxide-rich matrix

Table 4 Chemical composition of the Campo Bonito Formation; analyses by WDXRF, by D.F. Siems and T.R. Peacock.

	1	2
SiO_2	63.2	61.2
Al_2O_3	16.5	17.7
Fe_2O_3	6.89	7.46
FeO	0.68	0.48
MgO	0.88	0.90
CaO	0.07	0.13
Na_2O	<0.15	<0.15
K_2O	6.81	7.36
P_2O_5	0.09	0.11
TiO_2	1.14	1.16

Descriptions
1. Sample FT 762 (loc. 30), hematitic siltstone from Peppersauce Wash.
2. Sample FT 958 (loc. 29), hematitic siltstone from Southern Belle canyon.

Figure 31 Photomicrograph of maroon siltstone of the Campo Bonito Formation (loc. 30). Field of view 2 mm, plane light. Note the angularity and poor sorting of clasts of quartz, feldspar, and biotite, and the hematitic cement.

Figure 33 Diamictite unit of Campo Bonito Formation in the bed of Canada del Oro (loc. 32).

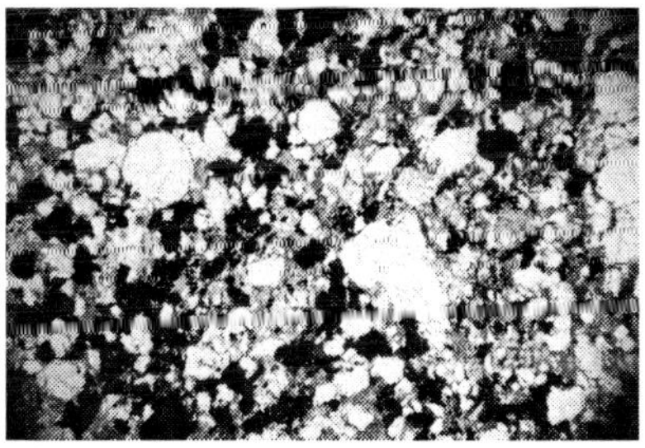

Figure 34 Photomicrograph of diamictite matrix; angular poorly sorted clasts include quartzite, granite, schist, volcanics, and jasper. Field of view 6 mm, plane light (loc. 32).

comprising 40 to 50 percent of the rock (fig. 31). Two specimens of the maroon hematitic siltstone were chemically analyzed (table 4). Ferric iron is unusually abundant, but the unit is not iron-formation. High potassium contents are consistent with abundant mica and potassium feldspar, but metasomatism is possible.

The maroon siltstone rests unconformably on upper Dripping Spring Quartzite and Proterozoic diabase (loc. 29). Regional erosion of Mescal Limestone (and perhaps Troy Quartzite, which occurs nearby) apparently took place on this unconformity, as the area of maroon siltstone is within the area of missing Mescal. The unconformity itself seems to be a preferred horizon in workings of the Southern Belle mine. Rocks below the unconformity locally appear weathered; in my opinion, Creasey (1967, p. 22) described the deeper parts of a paleosol as a peculiar oxide-mottled occurrence of uppermost Dripping Spring Quartzite in this area.

Diamictite. These rocks form a mappable unit in two areas on opposite sides of the Geesaman fault. The first area is in the Canada del Oro valley (pl. 1), where two well-exposed sections (locs. 32, 33) occur in the stream bed. The base of the unit is not exposed, but steep-sided paleo-channel walls can be inferred (fig. 32). Poorly sorted sedimentary breccia (diamictite) about 5 m thick is the lowermost lithotype observed (fig. 33), overlain by upward-fining laminated poorly-sorted sandstone with scattered angular granules and pebbles, mostly of a variety of rock fragments (fig. 34). Diamictite is overlain by Bolsa Quartzite.

The second area is in Atchley Canyon (loc. 20), where the diamictite is similar to that in Canada del Oro in lithology, sequence, and thickness. Conglomerate beds recur at the top of the unit. A high-angle channel wall cut in diabase is present but poorly exposed on the west side. The base of the diamictite in contact with Dripping Spring Quartzite is exposed at this locality.

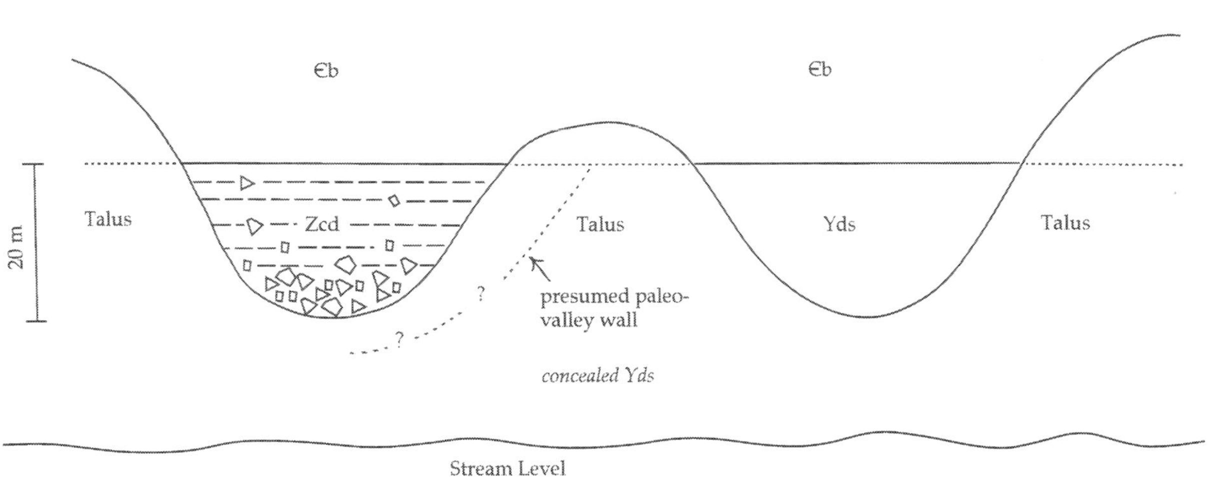

Figure 32 Sketch of relations among Dripping Spring Quartzite (Yds), diamictite of the Campo Bonito Formation (Zcd), and Bolsa Quartzite (Cb) by the bed of Canada del Oro (loc. 32).

Origin, age, and correlation. Correlation of the diamictite with the maroon siltstone is suggested by their identical stratigraphic position, their valley-fill morphology, their similarity in clastic petrology, and the presence of diamictite-like conglomerate at the base of the maroon siltstone. Both units are thought to be glaciomarine.

The maroon siltstone and diamictite could have been deposited any time between diabase intrusion (about 1100 Ma) and the Middle Cambrian (about 540 Ma). The lack of gradation between maroon siltstone and overlying Bolsa Quartzite, and lack of trace or body fossils in a hospitable lithology make a Precambrian age seem more likely. The lines of evidence suggesting a glaciomarine origin are:

1) High depositional relief, limited distribution, evidence from missing Mescal Limestone of local erosion, and valley-fill morphology.

2) Angular clasts, poor sorting, and mineralogical immaturity.

3) Association of diamictite with laminated (e.g. varved) fine-grained rocks.

These factors suggest glaciomarine deposition on a high-relief shoreline. Coupled with high iron content of maroon siltstone, they suggest a relation with Precambrian iron formation of the Rapitan type. These are about 700 to 800 Ma in age, and characteristically formed in glacial fjords (Maynard, 1991). Deposits of this type were not well known when Creasey (1967) published.

Glacigenic Precambrian rocks have not been previously recognized in southeastern Arizona, but glacial diamictites of this age have been recognized in Utah and southern California (Stewart, 1972). The nearest such locality is in southwestern New Mexico, about 250 km to the east (Corbitt and Woodward, 1981). The paleolatitude of the study area in the latest Precambrian was sufficiently high (Bond, 1984) to support glaciers.

The distribution of all outcrops of maroon siltstone and diamictite suggest the possibility of a single former valley open to the northeast, after reconstruction of the Geesaman fault. Since Mescal Limestone is missing everywhere the maroon siltstone and diamictite are present, and over a wide area toward the east and north, I presume that the valley was formed largely by erosion of the Mescal. The minimum depth of the valley was about 20 m, the approximate thickness of the maroon siltstone and diamictite, but a greater thickness of eroded Mescal probably represents some of the relief on the upper valley slopes.

Designation as Campo Bonito Formation. It is proposed that the marine siltstone subunit of the Bolsa Quartzite in the sense of Creasey (1967) be separated from that formation and, along with correlative diamictite, given formational status. The name proposed is the Campo Bonito Formation, from the former mining settlement.

The type locality proposed for the formation is in the canyon below the Southern Belle mine (loc. 29), where well exposed maroon siltstone reaches its maximum thickness of about 30 m and includes a diamictite-like basal layer. The age of the formation is believed to be Late Proterozoic.

PALEOZOIC SEDIMENTARY ROCKS

Paleozoic rocks of the study area form a sequence without angular discordance, as much as approximately 820 m (2689 ft) thick. The Paleozoic was largely a time of crustal stability and sediment accumulation in epicontinental seas in this region.

Figure 35 Typical appearance of cliffs formed by Bolsa Quartzite, showing upper light-colored more resistant part. A. Outcrop about 20 m high in Atchley Canyon (loc. 34); B. Distant view into Edgar Canyon looking northeast. Єa, Abrigo Formation; Єb, Bolsa Quartzite; Yds, Dripping Spring Formation.

Bolsa Quartzite

Distribution and description. From the Mogul fault to the range crest, the Bolsa Quartzite forms a widespread structural marker in every sizable structural block (pl. 1). However, its thickness varies greatly. Creasey (1967, p. 25) measured its thickness (as redefined in this study) as 93 m (305 ft), divided into an upper more massive, thick-bedded, coarse- to medium-grained part, and a lower part that is less resistant, medium-grained, thinner bedded, has a ferruginous matrix in some horizons, and is locally feldspathic. Waag (1968) measured three sections of Bolsa on the range crest as 21, 72, and over 43 m thick (70, 237, and 140 ft); Suemnicht (1977) obtained 104 m (340 ft) near Marble Peak. In Alder and Atchley Canyons, the thickness of the Bolsa varies laterally from about 12 to 92 m (40 to 300 ft) within distances of less than a kilometer (for example, west of locality 20).

The Bolsa shows a distinctive vertical sequence (Creasey, 1967); finer and less resistant maroon to gray beds toward the base pass upward into coarser and more resistant light-colored sandstone (fig. 35). This corresponds to a transition from festoon cross-beds upward into smaller herringbone cross-beds in some sections. Minor pebbly and gritty horizons are common toward the base. *Scolithus* tubes were noted by Creasey (1967) but I saw none.

Petrographically, the Bolsa consists almost entirely of quartz, with quartz grains showing interlocking quartz overgrowths. Locally the quartz is sutured, and metamorphic minerals may occur. Many beds are rather poorly sorted (Creasey, 1967) and probably bimodal.

Basal unconformity. Within the study area, the Bolsa rests disconformably on the Campo Bonito Formation near Peppersauce Wash (loc. 29), Canada del Oro, and Atchley Canyon (loc. 20), on Mescal Limestone near Mt. Bigelow (loc. 35), Edgar Canyon, and Canada del Oro, on the Dripping Spring Quartzite in many places, on Precambrian diabase in many places (loc. 34, 36), and on Pioneer Formation near Lemmon Rock (loc. 36). The Bolsa appears to be thinner where it rests on lower units (diabase and Pioneer in the Mt. Lemmon area; fig. 36). Thus the Bolsa must have been deposited on an irregular erosion surface cut on several rock units, with former topographic highs on former structural highs such as Mt. Lemmon. Such variation is common for the Bolsa (Middleton, 1989).

In the study area, the sub-Bolsa (including sub-Campo Bonito) unconformity has cut out the Troy Quartzite. Troy is absent in much of the region but present to the north (Krieger, 1968).

Age, origin, and correlation. The Bolsa Quartzite is thought to be Middle Cambrian, based on regional correlation. Probably the Bolsa is non-marine toward its base and marginal marine toward its top, with shoreline environments represented in the massive upper part.

In all the nearby ranges to the south, southeast, and southwest, Bolsa Quartzite forms the base of the Paleozoic section. To the north, however, it pinches out against Precambrian rocks (Krieger, 1968). Toward the east, the facies equivalent of the Bolsa, the Bliss Sandstone, is Upper Cambrian and Lower Ordovician (reviewed by Middleton, 1989).

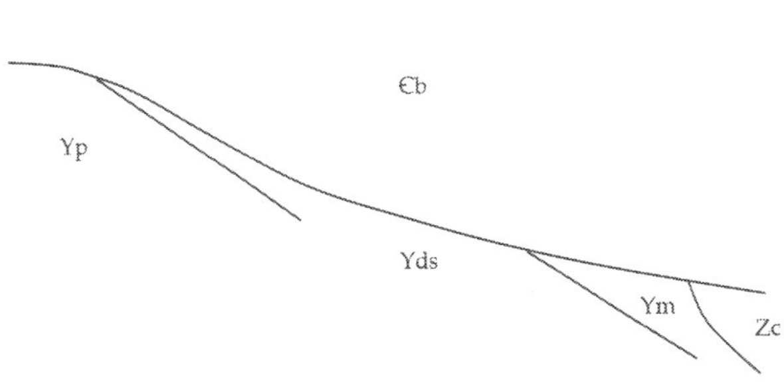

Figure 36 Diagram relating the thickness variations in Bolsa Quartzite to the stratigraphic level of its base. Not to scale; symbols as in plate 1.

Abrigo Formation

Finely bedded, fine-grained, carbonate-bearing non-resistant rocks conformably overlying Bolsa Quartzite belong to the Abrigo Formation. Creasey (1967) obtained a total thickness of 224 m (735 ft) in Nugget Canyon. Between Alder and Atchley Canyons, however, its thickness is locally less than 60 m (200 ft). The Abrigo Formation is extensively altered in the study area either by metamorphism or leaching. The least altered Abrigo encountered is in Alder Canyon (loc. 37).

Distribution. The distribution of Abrigo Formation is similar to that of the conformably underlying Bolsa Quartzite; that is, Abrigo is found in every major structural block between the Mogul fault and the range crest. Large areas of the main north-facing slopes of the range are Abrigo dip-slopes. Thus the Abrigo is distributed far more widely than some authors thought, and this has implications for mineral resource potential.

Three C Member. The basal member of the Abrigo consists of thin-bedded fine feldspathic sandstone with shaly partings and minor dolomite. Creasey

(1967, p. 29) obtained a thickness of 102 m (335 ft). As shown in his figure 3, the bedding could best be described as a somewhat flattened flaser bedding. A trilobite-brachiopod fauna was described by Stoyanow (1936) from Peppersauce canyon.

In the Mt. Lemmon area and just north of the range crest, this lower member of the Abrigo consists predominantly of calc-silicate rock with robust flaser bedding (fig. 37) but consistently includes impure calcite marble. The mineralogy is a sensitive function of metamorphic grade. Waag (1968) found that the lowermost lithotype of the member is a biotite-chlorite-quartz rock about 10 m thick.

Along the range crest, therefore, the Three C Member is considerably more calcic than in the Peppersauce Wash area studied by Creasey (1967). I think this reflects preservation of the calcic components in the high country by metamorphism; the calcium was baked in. In contrast, some component, probably calcium as calcite, was leached out in Creasey's area, as reflected by the earthy appearance noted by Creasey, the vuggy texture discernible in hand lens or thin section, and apparent flattening of sedimentary structures. Although the appearance of the rock suggests that leaching may have been by weathering, its rusty, vuggy nature extends below stream level (loc. 38), so leaching probably predates weathering.

Southern Belle Member. The middle member of the Abrigo Formation is a resistant white quartzite 5 to 6 m (15 to 20 ft) thick. Quartz grains are medium to coarse, well-rounded, and well-sorted, with interlocking overgrowths. Low-angle cross-lamination is characteristic. This member would be difficult to differentiate from Bolsa Quartzite were it not for the intervening Three C Member. The Southern Belle Member has been found wherever Abrigo Formation is present, from the Mogul fault southward to Mt. Lemmon, and eastward beyond the study area. It is shown only locally on plate 1, however.

On Mt. Lemmon, the Southern Belle Member is commonly a spotted hornfels. Waag (1968) found that the spots show calc-silicate assemblages. Probably the spots pseudomorph a nodular carbonate cement. Toward the top of the member, the place of the spots is taken by flattened vugs, suggesting that nodular cement was leached there before metamorphism.

Peppersauce Member. The upper member of the Abrigo is described by Creasey (1967) as dolomite and dolomitic sandstone, with a thickness of 102 m (335 ft) in the section presented on his page 33. The coarser upper portion shows cross-bed sets more than 30 cm high in dolomitic coarse sandstone; the finer lower portion shows flaser bedding and intraformational conglomerate. Stoyanow (1936) described a trilobite fauna from the base of the member in Peppersauce canyon.

In the Mt. Lemmon area, this member is mostly missing due to Paleozoic erosion, but it is well represented as calc-silicate rock similar to that of the lower member north of Mt. Bigelow. Like the lower member, the Peppersauce Member may have been leached of its calcite in the sections near Peppersauce Wash, as Creasey (1967) noted an increase in calcite in hornfelsed facies adjacent to Laramide porphyry. Leaching of calcite apparently postdated local fixing of calcium by metamorphism.

Age, origin, and correlation. The Abrigo is assigned a Middle to Late Cambrian age, based on its contained trilobite faunas. The flaser-bedded portions of the Abrigo were deposited in intertidal environments. The combination of leaching in the Peppersauce Wash-Nugget Canyon area, and metamorphism elsewhere, makes the Catalinas a poor range in which to study sedimentary features of the Abrigo, however.

The middle quartzite member probably reflects a shoreline environment. Its persistence across the study area suggests that it formed due to some widespread change of conditions. Epis and Gilbert (1957) suggested that the middle member thickens regionally eastward and is contiguous with the Upper Cambrian and Lower Ordovician Bliss Sandstone of New Mexico. This geometry is similar to that of a transgressive-regressive wedge (as noted by Hayes and Landis, 1965), but if so the depositional relation between intertidal carbonates and shoreline sands is unclear.

The distribution of the Abrigo in nearby ranges is the same as that of the Bolsa. Toward the east, the facies

Figure 37 Specimen showing flaser bedding in Abrigo Formation, converted to calc-silicate mineral banding by metamorphism, from Butterfly Peak (loc. 35).

equivalent of the Abrigo is generally considered the El Paso Formation of Late Cambrian and Early Ordovician age (Hayes, 1972; reviewed by Middleton, 1989).

Mt. Lemmon Unit (informal)

Waag (1968) has described the uppermost two units of the Abrigo Formation in the Mt. Lemmon area as a quartzite and phyllite that do not contain any calcic layers. These units form a chain of hilltops and ridges (pl. 1), among which is Mt. Lemmon, the highest point in the study area. I think that they are not part of the Abrigo Formation, but constitute a younger fluvial unit.

Distribution and description. The fluvial unit shows a linear distribution that extends from Mt. Lemmon down Red Ridge (pl. 1). Preservation of the unit is partly fortuitous, depending on the intersection of the gently NE-dipping but deformed fluvial unit, with the high-relief NE-sloping land surface. To some extent the unit has apparently controlled the topography, however, so that its outcrop pattern probably reflects a linear original distribution. Nowhere are overlying units preserved.

The fluvial unit, approximately 50 m (160 ft) thick, consists of a thin basal conglomerate containing white quartzite clasts, overlain by a sequence of amalgamated micaceous sandstone beds separated by flaggy arenaceous phyllite layers (loc. 39), all intruded by later pegmatite. Cut-and-fill structures are characteristic of the sands but directional features are obscure. The thicker sandstone beds are poorly sorted toward the base of the unit, and sorting improves upward. Isolated granular or pebbly horizons occur throughout the unit. The phyllitic layers are commonly hematitic, and contain calc-silicate assemblages in nodular domains about 10 by 20 cm, probably of carbonate originally. Locally, graphitic beds of 2 to 10 cm thickness occur; these must originally have been coals.

Basal unconformity. The fluvial unit has been cut into the Abrigo Formation; a basal conglomerate is locally present. Disconformity is locally visible at outcrop scale, but complicated by deformation. Most commonly, part or all of the upper (Peppersauce) member of the Abrigo Formation is missing where the fluvial unit is present. Along Red Ridge (loc. 40), the fluvial unit rests on the middle quartzite (Southern Belle) member for the entire length where the fluvial unit is present, about a kilometer. On Mt. Lemmon, the flattened vugs at the top of the middle member of the Abrigo are probably related to dissolution of nodular cement along the unconformity. At the top of the ski lift, a small thickness of the upper member of the Abrigo remains under the fluvial unit.

Age, origin, and correlation. The fluvial unit seems to be a channel fill, based on its shoestring distribution and assemblage of sedimentary features. The channel was cut in a substrate of Abrigo Formation.

This origin has implications for the age of the unit. Regionally the Abrigo Formation is separated from the overlying Martin Formation of marine origin by a disconformity that spans Ordovician through Middle Devonian time. The fluvial unit seems far more likely to belong above this disconformity than below it. That is, the fluvial unit would be the fluvial record of erosion on this unconformity that initiated the Martin depositional interval. This in turn implies that the fluvial unit is Devonian in age. The presence of paleo-coals is consistent with a Devonian age.

No similar units have been reported in the immediate area, but fluvial sands form the base of the Martin Formation in the Payson area to the north (Shride, 1967; Krieger, 1968). There they are Lower or Middle Devonian (Teichert and Schopf, 1958).

Designation as Mt. Lemmon unit. It is proposed that the fluvial unit be informally designated the Mt. Lemmon unit. The excellent exposures on the road near locality 39, which expose the maximum thickness of the formation, are designated type section. Formal description awaits documentation of wider distribution. An alternate reference section (loc. 40) is along the crest of Red Ridge, where the formation rests on the middle quartzite (Southern Belle) member of the Abrigo Formation. This section is more deformed and recrystallized. In no section is the top of the formation exposed. The character and thickness of the formation is described above. The age is thought to be Devonian, probably Early and(or) Middle Devonian.

Martin Formation

Distribution and description. Disconformably overlying the Abrigo Formation between the Mogul and Edgar faults is the dolomitic Martin Formation. The distribution of Martin is quite regular in this area. South of the Edgar fault, the Martin has been eroded. Thus no contact is exposed between the Martin and the Mt. Lemmon unit described above.

The Martin Formation is 77 m (252 ft) thick (Creasey, 1967, p. 36), and consists toward the base of mostly gray dolomite, and toward the top of yellow-weathering dolomite, separated and capped by dolomitic sandstone intervals. Thinly-bedded, bioturbated, impure dolomite forms a basal lithotype in some sections. Medium-grained dolomite in beds 0.3 to 0.5 m thick is the most characteristic carbonate rock type, but limestone is present. *Atrypa* and other brachiopods are abundant in some beds, corals, *Camarotechia*, or fish teeth in others (Stoyanow, 1936, p. 489). Hematitic oolite horizons (the O'Carroll beds of some authors) were not recognized in the study area.

Based on conodont biostratigraphy, some authors have proposed that the upper clastic-bearing parts of the Upper Devonian be separated as the Percha Formation. Schumaker (1978) used Stoyanow's (1936) section to suggest a local thickness of 46 m for the Percha. I was not able to separate this unit for plate 1.

The Martin-Escabrosa contact coincides with the Devonian-Mississippian boundary. This puts the boundary within dolomitic beds with slightly different weathering character, and makes the contact difficult to map. In metamorphosed sections such as at Marble Peak, the presence or absence of Martin is difficult to establish. I have chosen to show Martin in such sections on plate 2B by assigning most of the coarsely crystalline dolomite in rough-textured outcrops to the Martin.

The great disconformity at the base of the Martin, representing the Ordovician, Silurian, and part of the Devonian, is not obvious in outcrop. This disconformity is thought to be the same as that described at the base of the Mt. Lemmon unit.

Age, origin, and correlation. The age of the Martin (and Percha) Formations is Late Devonian, though the uppermost Devonian is not represented. The Martin is more widespread than the lower Paleozoic formations, as it is present not only in the ranges to the south, but also those to the north and east, where it overlies Precambrian rocks. The Late Devonian was a time of high sea level, and regional relations suggest that the Martin may be a transgressive epicontinental-sea deposit.

Escabrosa Limestone

Among the carbonate units of the study area, the Escabrosa Limestone is the most resistant; it commonly forms a curtain of gray cliffs.

Distribution and description. Between the Mogul and Geesaman faults, the Escabrosa is the uppermost Paleozoic sedimentary rock, as it dips northeast into a complex contact with American Flag Formation. South of the Geesaman fault, Escabrosa is present in a more complete Paleozoic section as on Marble Peak. South of the Edgar fault, it has been removed by erosion. The Escabrosa apparently rests conformably on underlying Martin Formation.

The Escabrosa is characteristically a well-sorted calcarenite in thick beds. Its base is dolomitic. Bedding lamination is most conspicuous near the top of the formation. Near Marble Peak, a thin greenish metachert-pebble conglomerate serves as a marker horizon within the Escabrosa. Nodules of chert occur sporadically through the formation. A section measured by Creasey (1967, p. 39) totals 166 m (545 ft) but does not include the top of the formation. Crinoidal debris is an important component, and rugose corals including the colonial coral *Lithostrotionella* are common. Bryozoan debris is locally abundant.

The Escabrosa is intruded by Leatherwood granodiorite south of the Geesaman fault. Near the intrusive contacts the Escabrosa is recrystallized to coarse white marble (fig. 38), except for the basal interval which shows coarse dolomite rhombs.

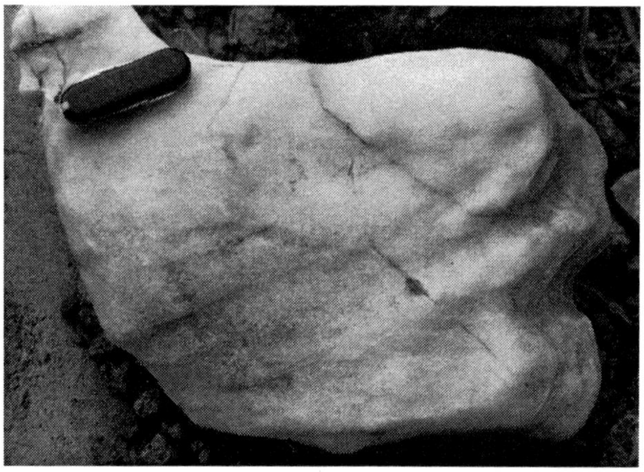

Figure 38 White marble developed from Escabrosa Limestone near Leatherwood granodiorite intrusion, Geesaman Wash (loc. 41).

Figure 39 Folds (outlined) in Horquilla Limestone at Marble Peak. Outcrop about 1.5 m high.

Figure 40 Some sedimentary and diagenetic structures in the Horquilla Limestone at Marble Peak. A. Anastomosing dolomite after flaser bedding in the Horquilla Limestone at Marble Peak (loc. 42); B. Chert-dolomite-phosphate nodules in laminar limestone (loc. 43).

Figure 41 Eastern termination of the Leatherwood (Kl) sill into Earp Formation (Pe) quartzite, ridge east of Lombar Peak. PPh, Horquilla Formation.

The Escabrosa forms caves more commonly than the other carbonate units of the study area. The orientation of the Peppersauce Cave (John Priser in Creasey, 1967, fig. 6) suggests that it formed after mid-Tertiary tilting and marks a higher water table than the present one; that is, the cave is developed along the strike of the NNE-dipping unit with a few shoots down the dip.

Age, origin, and correlation. The Escabrosa is the local record of extensive Lower and Upper Mississippian epicontinental seas. It correlates with the Redwall Limestone of the Colorado Plateau. To the southeast and southwest, the Escabrosa is found in many ranges to the Mexican border. Just southeast of the study area, the Escabrosa is thinner than elsewhere in the region (Peirce, 1979).

Naco Group

Distribution and description. The Middle Pennsylvanian to Lower Permian Naco Group has a restricted distribution in the study area. However, the group is quite well exposed in two small areas.

The two areas of Naco Group rocks are between the Geesaman and Edgar faults and are separated from one another by the voluminous Leatherwood granodiorite sheet. North of the Geesaman fault, Laramide erosion removed the Naco prior to venting of the Rice Peak porphyry and deposition of the American Flag Formation. South of the Edgar fault, the Naco has been removed by modern erosion. Probably it was originally present in both blocks.

The most extensive area of Naco Group is that between Geesaman and Alder washes, below the Leatherwood intrusive sheet. Here the Naco is moderately to severely deformed, and toward the western end extensively recrystallized. This area has small outliers isolated by erosion south of Alder Creek.

The second area, to the west, is on the upper slopes of Marble Peak above the Leatherwood sheet, where the group is strongly recrystallized and badly deformed (fig. 39). Some sedimentary and diagenetic features are nevertheless well preserved (fig. 40).

The Naco Group stratigraphy recorded in the two areas is similar, and in accordance with that described for the immediate region by Blakey and Knepp (1989). Horquilla Limestone at the base is overlain by an interval of interfingering limestone, argillite, and fine white quartzite assigned to the Earp Formation. Generally, most of the Earp consists of quartzite (fig. 41). No unconformity at the base of the Naco Group was observed in the study area. However, Lower Pennsylvanian rocks are not present, and Black Prince Limestone of Early Pennsylvanian age is present about 50 km to the southeast (Bryant, 1968), so a local disconformity representing that time period seems likely.

Horquilla Limestone. In both areas, the Horquilla consists of finely bedded carbonate rocks. It is best exposed in the Marble Peak area, where the basal interval consists of limestone intricately replaced by dolomite to form a distinctive pattern (fig. 40A) controlled mostly by bedding. The upper interval consists of finely laminated dark limestone, one horizon of which contains dolomite-chert-phosphate nodules (fig. 40B). The total apparent thickness is about 180 m (590 ft). Isoclinal and disharmonic folding (fig. 39) are ubiquitous, but the Escabrosa-Horquilla contact is a gently warped plane at map scale.

One fossil locality was found in dolomitic lower Horquilla on Marble Peak (loc. 42). It yielded a rhynchonellid brachiopod, crinoid calyx (?) and columnals, and rugose corals. Preservation is poor, and no fusilinids were seen.

Earp Formation. Overlying Horquilla Limestone is an interval about 100 m (340 ft) thick consisting of interfingering limestone, argillite, and quartzite (fig. 42). The limestone is much like that of the upper Horquilla. The quartzite, as much as 80 m thick, is fine, commonly silicified, and finely bedded. The interfingering is apparently sedimentary in part, though clearly it could be accentuated by folding. The thickness is a minimum because the original top is nowhere exposed. On Marble Peak, quartzite forms the southern of two peaks of nearly equal height, but an overlying limestone forms the other. The resistance to erosion of this quartzite is probably responsible for the existence of Marble Peak.

In the eastern area, the base of the quartzite is a chert-pebble, limestone-matrix conglomerate (loc. 45), apparently the "jelly-bean" conglomerate characteristic of the Earp Formation in other areas. Janecke (1986, her fig. 4; 1987) mistakenly referred this conglomerate to the Bisbee Group.

Age, origin, and correlation. The parts of the Naco Group present in the study area represent Middle Pennsylvanian (Atokan or Desmoinesian Series) through earliest Permian time. More complete sections of the Naco Group are available in many of the ranges in all directions around the study area. The general depositional environment of the Naco Group in this area was a subtidal shelf, marginal to the Pedregosa Basin to the southeast, except during the Pennsylvanian-Permian boundary interval, when the Earp Formation was deposited in a partly intertidal distal margin of a delta to the northwest (Ross, 1978).

The dolomitic facies of the Horquilla, its modest thickness, and the clastic facies of the Earp suggest that the study area formed an interbasinal high. This in turn suggests that the disharmonic folding shown by the Naco Group in the study area is probably not a primary (slump folding) feature.

SEDIMENTARY AND VOLCANIC ROCKS OF MESOZOIC AND TERTIARY AGE

Unlike the Paleozoic, accumulation in the study area of sediments and associated volcanics during the Mesozoic and Tertiary was a response to tectonism, which can be divided chronologically into 1) probable Early Cretaceous and Late Jurassic crustal rifting, leading to accumulation of the Bisbee Group, 2) Late Cretaceous to early Tertiary (Laramide) shortening and uplift, which led to accumulation of the American Flag Formation, and 3) mid-Tertiary crustal stretching and corollary tilting, leading to accumulation of several volcanic and coarse clastic units. Lower Mesozoic rocks representing a total of about 80 million years are missing from the study area; rocks in the upper part of this missing interval are reported about 40 km to the south.

Probable Bisbee Group

Janecke (1986, 1987) mapped the Bisbee Group in a small area on the south (hanging-) wall of the Geesaman fault. On plate 1, I show probable Bisbee Group there in an even smaller area, because one unit that Janecke included in Bisbee Group is actually part of the Naco Group, and the presence of Bisbee Group is

Figure 42 Interfingering of marble (m, above), quartzite (q), and sandy argillite (a, below) in Earp Formation, Marble Peak (loc. 44).

not based on strong evidence. This evidence merely consists of permissive lithology, stratigraphic position on Naco Group, and association with an old fault.

The probable Bisbee Group rocks occupy 0.1 km^2 or less (pl. 1), south of the Geesaman fault, adjacent to the San Pedro basin on the eastern edge of the study area. The rocks most confidently assigned to Bisbee Group are a quartzite and a local polymictic conglomerate at the base. The thickness is less than 100 m in the apparently north-dipping sequence truncated on the north by the Geesaman fault. Possibly some maroon to green phyllite under the conglomerate also belongs to the Bisbee Group, extending the possible thickness to about 150 m. However, the Bisbee Group does not include a basal limestone-pebble conglomerate as Janecke (1986, her fig. 4; my locality 45) stated. Mediocre exposure and a pervasive south-dipping cleavage prevent detailed description.

The Bisbee Group is currently thought to include Upper Jurassic (Kluth, 1982; Bilodeau and others, 1987) as well as Lower Cretaceous rocks. The group is widespread in southeastern Arizona but fills discrete rift basins elongate northwest-southeast, in response to concurrent rifting on the same trend (Bilodeau, 1982). Marine facies in the broad Chihuahua trough are common to the southeast (Dickinson and others, 1986). Bisbee Group occurrences nearest the study area, 16 km to the southeast, are associated with faults that are down to the southwest. Janecke (1986) suggested such a relation with the Geesaman fault for the Bisbee Group in the study area.

American Flag Formation

Though the American Flag Formation has very limited distribution, its potential is great as a recorder of Laramide events in the study area. This potential is only partially realized in this paper; more detailed work is warranted.

I conclude that American Flag Formation records three different stages in the Laramide history of the area. I have not divided it formally, pending more information, especially on the age of each subunit.

Distribution and general description. American Flag Formation is present in two areas about a kilometer apart on the northeastern margin of northeast-facing bedrock sequences between the Geesaman and Mogul faults. The first is between the Mogul fault and Peppersauce Wash in the area of 3C Ranch. The second area south of Peppersauce Wash is better exposed and more extensive (pl. 2A), and my description of the parts of the formation is based on this second area.

The formation was introduced by Creasey (1967), who described it as green gray graywackes and conglomerates, intruded by granodiorite (Rice Peak) porphyry. Janecke (1986, 1987), however, presented evidence for unconformable relations between American Flag Formation and underlying Rice Peak porphyry. Creasey assigned a Cretaceous (?) age to the American Flag, based in part on a non-marine fauna reported by Bromfield (1950), and included a stratigraphic section from Nugget Canyon. The total listed thickness is 2080 ft (635 m) but the top is not exposed.

It appears to me that seeming contradictions in the descriptions of the American Flag Formation are resolved by considering it as three sub-units that responded to rapidly evolving structural and intrusive conditions. These are 1) a lower member that accumulated in response to northeastward movement of the Paleozoic sequence on reverse faults, under which the member is preserved; this member is intruded by Rice Peak porphyry and contains mostly limestone clasts; 2) a middle member that accumulated after reverse faulting, during unroofing of Precambrian rocks, and is unconformable on Paleozoic rocks; it is intruded by one type of Rice Peak porphyry, but contains a few clasts of another type; and 3) an upper member not intruded by Rice Peak at the level of exposure but affected by hydrothermal activity, and containing abundant Rice Peak clasts. The stratigraphy of American Flag Formation in this view is thus inseparable from structural and igneous events, and serves as a recorder of them. Correlation between the two areas of American Flag Formation is aided by reference to position in this evolutionary sequence.

The three informal members are shown in plate 2A. Relations of the three members are shown in the cross section. Note that the top of the formation is not exposed. Subdivision of Rice Peak porphyry (described below) reinforces the proposed sequence.

Lower member. Limestone-cobble conglomerate characterizes the lower member. Interbedded lithotypes are flinty phyllite and sandstone, feldspathic toward the top, commonly with isolated limestone cobbles. The lithologic assemblage and complex distribution relative to Paleozoic rocks suggests a paleokarst aspect. Vertebrate and invertebrate fossils have been found in the lower member by Paul Olsen and Kathryn Gregory (oral commun., 1997).

The member includes approximately beds 1 through 5 (557 ft or 170 m thick) in the Nugget Canyon section measured by Creasey (1967, p. 44); this section, measured in a canyon, is the most complete (cf. pl. 2A). The fine-grained basal beds described by Creasey are supplanted by conglomerate in other sections, and grade laterally to conglomerate in this one.

Beds of the lower member are commonly steeply dipping, locally overturned to the northeast, and contain a NW-dipping cleavage. The structural base is locally formed by reverse faults that put Escabrosa Limestone over American Flag, but elsewhere conglomerate of the lower member rests directly on Paleozoic rocks. The relation of faults to the basal member is most conveniently exposed in two areas within the Peppersauce-Nugget Canyon outcrop belt—the first south of Nugget Canyon (loc. 46) and the second at the northern end of the outcrop belt (loc. 47).

Porphyry bodies locally cut both the lower member (loc. 47) and the bounding faults. Alteration is to chlorite-epidote assemblages but is not conspicuous. In addition to limestone clasts, volcanic and quartzite clasts are present toward the top.

Middle member. Conglomerates containing clasts of Oracle Granite, Barnes-type conglomerate, and quartzite typify the middle member, along with gritty feldspathic poorly-sorted greenish sandstone. These rocks are commonly very tough due to a volcaniclastic matrix partially recrystallized to epidote and chlorite (loc. 48). A horizon of pale vitric tuff is present (loc. 49). This member may be as thick as 200 m locally.

The middle member unconformably overlies the package consisting of the lower member, Paleozoic rocks, and the reverse faults separating them. It forms gently northeast-dipping flatirons extending southwestward into the range along ridge tops (pl. 2A). The basal conglomerate beds contain limestone cobbles. The unconformity between the middle member and the underlying package is exposed at locality 50. Toward the northeast, this contact becomes conformable and gradational over about 5 m. I have found plant stem fragments in basal beds of the middle member at the southern end of the outcrop belt (loc. 51).

Small porphyry bodies are intrusive in the middle member (loc. 52), and a larger porphyry body was emplaced along the Paleozoic unconformity just north of Nugget Canyon (pl. 2A). The southern end of the outcrop belt is apparently formed by an intrusive contact with Rice Peak porphyry (loc. 51). However, sparse porphyry clasts are also present in conglomerates of the middle member. The varieties of Rice Peak porphyry involved in these various relations are described under that unit.

Upper member. Abundant porphyry clasts characterize the upper member. Conglomerate, the predominant rock type, contains clasts of Precambrian, sandstone, and volcanic rocks in addition to porphyry. One distinctive bed contains welded tuff clasts only (loc. 53). No intrusive porphyry bodies were seen. The upper member appears to rest conformably on the middle member in all sections, and the contact may be gradational over about 10 m.

The upper beds measured by Creasey (1967) in Nugget Canyon can not be partitioned between the middle and upper members based on his information. Probably the exposed thickness of the upper member is only about 100 m; however, the top of the member is covered.

Northern area. The American Flag Formation immediately adjacent to the Mogul fault contains rocks not exactly equivalent to those described above. All contacts with Paleozoic rocks there are reverse faults, adjacent to which is a sedimentary breccia about 10 to 15 m thick composed of Paleozoic and volcanic blocks. This is an apparent equivalent of the lower member. Overlying this basal lithology is reddened, altered and veined volcaniclastic sandstone and fine conglomerate with silicified limestone and volcanic pebbles, which is partitioned among the upper members of the southern area with difficulty.

Age, origin, and correlation. The age of the American Flag Formation is not known from direct evidence except the Cretaceous (?) plants. From relations to faults and intrusion it seems likely to be broadly Laramide. A Late Cretaceous age of the lower member is consistent with its relations to intrusive porphyry. The three members seem to record various stages in a single orogenic sequence. Thus the total age span of the formation may be small. The top of the formation is not exposed, so the total thickness and the age of oldest and youngest parts are unconstrained. A minimum age for the uppermost part is implied by its presence in the footwall sequence of a mid-Tertiary detachment fault. Creasey's (1967) 635 m thickness is the maximum for the study area but is commensurate with those of similar sequences in the region. Cross-sectional relations (pl. 1) suggest that a thickness as great as 910 m, still limited by structures and (or) later erosion, is plausible.

My descriptions of the formation have already implied an origin and geologic evolution of the formation with respect to deformation, source, intrusion, and alteration. They also imply that the American Flag Formation accumulated along the northeastern margin of a Laramide uplift that exposed Oracle Granite basement, overlying Apache Group, and Paleozoic sedimentary rocks (but not Wilderness suite granites), adjacent to a proto-San Pedro ?foreland basin, concurrent with volcanism and the earliest stages of Laramide plutonism. The depositional environment seems likely to have been on alluvial fans for most of the formation. Conglomerate-argillite relations in the basal member suggest accumulation from debris flows into a basin with poorly integrated drainage.

Laramide events in southeastern Arizona are thought by most authors (cf. Davis, 1979) to have been focused on discrete areas. If so, detailed correlation of the sedimentary responses between different areas is unlikely. As we have seen, the character of the American Flag Formation is strongly influenced by the behavior of immediately adjacent bedrock. In these circumstances, the only correlations that are tempting are those with sequences that could have been in the same basin or other flanks of the same range. These include the Cascabel Formation and Muleshoe Volcanics and the Glory Hole Volcanics on the San Pedro valley margin of the southern Galiuro Mountains (fig. 2), as the San Pedro area may have already been a single sedimentary basin or basin margin in Laramide time. The Cascabel-Muleshoe sequence (Goodlin and Mark, 1987), about 1750 m thick, consists of alluvial fan detritus derived from the north, and underlying intermediate volcanic rocks that span at least the period 75 to 66 Ma. The Glory Hole Volcanics of intermediate composition, about 500 m thick, accumulated at about 70 Ma in a location that could have been less than 20 km to the east from the American Flag depocenter during the Laramide, i.e., before mid-Tertiary extension. About 70 km north of the study area, Upper Cretaceous transgressive marine deposits of the Aravaipa-Winkelman area probably predate the American Flag Formation.

Probable American Flag Formation near San Manuel

In the south fork of Mammoth Wash in the San Manuel district (loc. 55), conglomerate and volcanic rocks previously mapped as Cloudburst Formation (Creasey, 1965) are intruded by a large porphyry dike. Unruh in this volume shows this porphyry to be about 68 Ma in age. Therefore correlation of the dike with San Manuel porphyry, and its host rocks with American Flag Formation, seems appropriate. These hosts consist of clastic rocks, mostly conglomerate with cobbles of Oracle Granite and volcanic rocks, overlain by amygdaloidal intermediate volcanics and flow-breccias. Dip is to the northeast.

The presence of a Laramide supracrustal sequence in the San Manuel district is significant for that district, regardless of correlation nomenclature. On plate 1, this sequence is shown as probable American Flag Formation; its age is pre-Tertiary based on Unruh's dates on intrusives in it (this volume), and is probably Late Cretaceous.

Mid- and Upper-Tertiary Units

Units of mid- to late Tertiary age form the two ends of the study area and one more central area. They are coarse detritus of non-marine origin, formed in response to multiple stages of crustal extension and tilting, along with coeval volcanics. Thus the older members of the sequence can be steeply dipping, and they are locally indurated and altered. Dickinson (1991) has reviewed the geology of Tertiary strata associated with all the faulted margins of the Catalinas. Force and Cox (1992), Dickinson (1993), Force and Dickinson (1994), and Force and others (1995) have provided recent descriptions of sequences on the northern margin, so they will not be described in detail here. The Tertiary sequences will be discussed primarily by area rather than age.

The Tertiary rocks on the northern margin of the study area will be characterized first. They are (from oldest to youngest) the Cloudburst Formation of late Oligocene and early Miocene age, the San Manuel Formation of early to middle Miocene age, and the Quiburis Formation of late Miocene to Pliocene age. The formations were first described by Heindl (1963), though the term Cloudburst was introduced by Pelletier (1957).

Cloudburst Formation. This unit of coarse clastic and volcanic rocks reaches its greatest thickness, about 3500 m, north of the Turtle fault and above the Cloudburst detachment fault (pl. 1). It dips steeply to the east and northeast in this area, and is locally indurated and altered. The formation is divided into a lower member consisting largely of intermediate and rhyolitic volcanic rocks with lesser conglomerate, and an upper member of reddish conglomerate with lesser rhyolitic tuff, derived toward the base from volcanic rocks but toward the top from Oracle Granite, porphyry, and diabase. Clast transport was largely from the west (Weibel, 1981), but some volcanics of the lower member may have been derived from the east (Force and Cox, 1992). Creasey (1965) has mapped some sedimentary breccia horizons derived from granitic rocks north of the map area.

Depositional trends suggest that accumulation was in response to movement on the Cloudburst-Turtle fault system, and south of that system only thin sequences lapped onto Oracle Granite basement (locs, 26, 57). Isotopic ages are 28.3 and 27.8 Ma toward the base and 22.5 Ma toward the top of the formation (Dickinson and

Shafiquallah, 1989; S. B. Keith and W. A. Rehrig, written commun.,1977). Chemical analyses are given in table 5; the volcanics (except rhyolite) are latitic, and the analyzed conglomerates partly reflect the abundance of clasts of Oracle Granite.

San Manuel Formation. This conglomeratic unit, as thick as 2000 m, overlies the Cloudburst Formation, unconformably in places, and is less-tilted, less-indurated, unaltered, and paler in matrix color. It is divided into a lower (Kannally) member derived from Oracle Granite and other rocks to the west, and an upper (Tucson Wash) member derived from Cloudburst conglomerate and volcanic rocks to the east. The Kannally Member contains sedimentary breccia bodies (Creasey, 1965) that apparently represent rock avalanches derived from Precambrian and Paleozoic rocks. Accumulation was mostly in response to movement on the San Manuel fault system (pl. 1), and two splays of the fault may show differing histories of accumulation of San Manuel Formation (Force and Cox, 1992). Radiometric ages of included tuffs span the interval 22 to 16 Ma (Dickinson and Shafiqullah, 1989).

Together the Cloudburst and San Manuel Formations form an indurated to semi-consolidated, variably deformed western wall of the San Pedro basin. They are quasi-bedrock for late Tertiary events.

Quiburis Formation. This gravel unit fills the modern San Pedro basinal trough; dips are less than 10° to the northeast. It accumulated in response to movements on basin-and-range faults such as the Mammoth and Cholla faults (pl. 1). The formation consists mostly of alluvial fan deposits but contains a lacustrine facies (east of the study area) with associated gypsum and diatomite deposits. Its maximum thickness is only about 500 m, and some older features are not completely covered. Quiburis Formation appears to be cut by the youngest stage of movement on the Mogul fault. Its age may range from mid-Miocene through Pliocene.

Stratton unit (informal). In the bed of Stratton Wash and on the hillslope to the northeast (loc. 58), surrounded by alluvium at the edge of the San Pedro basin, is a northeast-dipping sequence that consists from the base of 1) maroon to greenish conglomerate and poorly sorted gritty sand, containing angular clasts of volcanic rocks and sedimentary rocks like those of the American Flag Formation, 2) vesicular flows of intermediate composition and microlitic texture, and 3) coarsely porphyritic gray volcanic rock, locally flow-banded, with plagioclase, quartz, biotite, and amphibole phenocrysts. The exposed thickness is 200 to 300 m (650 to 1000 ft), most of which is in the volcanic rocks. In the best exposure at locality 58, neither the top nor the base of the unit is exposed; the top is eroded and(or) covered, and the Stratton detachment fault separates the unit from older adjacent rocks. Cross section A-A' (pl. 1) suggests thicknesses of between 600 and 3000 m (2000 to 10,000 ft).

M. Shafiqullah (in Janecke, 1986) dated biotite from the upper volcanic interval by the K-Ar method. The age obtained was 27.5 Ma; uncertainty was not listed.

To the north in the bed of Peppersauce Wash (loc. 59), similar rocks are found between two splays of the Mogul fault. Though the rocks are shattered, approximately the same sequence as in Stratton Wash is present. Exposures to the southeast (pl. 1) extend the sequence upward and include more conglomerate.

The informal term Stratton unit is proposed for this sequence, and the admittedly incomplete section at locality 58 proposed for the type section. The age of the unit is apparently late Oligocene; it probably formed in alluvial fans and accumulated during detachment faulting.

Table 5 Chemical composition of conglomerate and volcanic rocks of the Cloudburst Formation and associated rhyolite and rhyodacite; * , all iron reported as one oxide.

	Rhyodacite instrusive	Rhyolite		Intermediate volcanic flows		Conglomerate
	1	2	3	4	5	6
SiO_2	66.7	77.2	n.d.	53.7	n.d	n.d.
Al_2O_3	16.2	12.9	12.3	15.8	16.1	12.1
Fe_2O_3	3.2	0.56	0.6	6.5	8.9	3.4
FeO	0.11	0.05	*	0.88	*	*
MgO	0.60	0.20	0.15	5.4	4.6	1.3
CaO	1.7	0.46	0.46	4.0	5.3	0.9
Na_2O	4.4	2.7	2.6	3.2	3.2	1.7
K_2O	4.4	5.1	4.9	4.6	5.5	5.9
TiO_2	0.64	0.1	0.1	0.93	1.4	0.5
P_2O_5	0.20	0.02	0.05	0.32	0.6	0.2

Descriptions

1. Intrusive in Purcell window, rapid rock analysis reported by Creasey (1965)
2. Rhyolite north of study area, rapid rock analysis reported by Creasey (1965)
3. Intrusive sericitic rhyolite from Tucson Wash (below loc. 94), ICP-AES analysis by P. H. Briggs.
4. Latite flow from Tar Wash, rapid rock analysis reported by Creasey (1965)
5. Intermediate flow near mouth of Cloudburst Wash (below loc. 56), ICP-AES analysis by P. H. Briggs.
6. Average of two conglomerates of lower member, both unweathered and containing Oracle Granite clasts, 10 to 30 m from Mammoth vein, samples FT132X (loc. 56) and 751 (loc. 57); ICP-AES analyses by P. H. Briggs.

Pantano Formation. The Tertiary rocks occupying the Tucson basin on the southern margin of the study area are (from the base) the Pantano Formation and the Tinaja beds. The stratigraphic terminology used here is after Davidson (1973) and Anderson (1987); an older terminology introduced by Pashley (1966) is superseded.

Bedrock of the Catalina Mountains is in detachment-fault contact (where exposed) with the Pantano Formation. Exposure of Pantano is intermittent (pl. 1); it is apparently deformed into irregular folds based on variable attitudes. The Pantano consists of boulder- to pebble-gravel with clasts from sources structurally high in the Catalinas. Elsewhere, the formation contains lacustrine facies, and interlayered intermediate volcanic flows dated as 24 to more than 28 Ma. Thus the Pantano is thought to be largely Oligocene. Toward the center of the Tucson basin, equivalents of the Pantano Formation are 1440 m (4700 ft) thick (Eberly and Stanley, 1978).

Tinaja beds. These gravels are commonly faulted against Pantano Formation, but elsewhere overlie it. The Tinaja within the map area consists largely of boulder- to pebble-gravel in beds with appreciable dip, commonly to the south. Clast type shows a transition within the unit; lower beds contain clasts from sources structurally high in the Catalinas, whereas higher beds have clasts such as mylonitized Wilderness suite leucogranites representing low structural positions (Voelger, 1953; Pashley, 1966). Thus the unit records uplift and unroofing of the southern part of the Catalinas. In other areas, the Tinaja beds include lacustrine facies, and volcanic rocks as young as 12 Ma. Thus the unit is thought to be mostly Miocene. It is about 2100 m (6900 ft) thick in the central Tucson basin (Eberly and Stanley, 1978). It has been subdivided by Pashley (1966) and Anderson (1987).

Figure 43 Mylonitic Leatherwood granodiorite at Molino Basin (loc. 60). Light-colored bands are younger Wilderness suite pegmatite and leucogranite. Field of view about 4 x 7 m.

PLUTONIC IGNEOUS ROCKS OF MESOZOIC AND TERTIARY AGE

The more extensive plutonic rock units range in age from Late Cretaceous to mid-Tertiary, and occur in three age groups, 1) Late Cretaceous to early Tertiary rocks including Leatherwood granodiorite, porphyries of Rice Peak, and porphyries of San Manuel, 2) Eocene leucogranites of the Wilderness suite, and 3) mid-Tertiary plutons of which the most voluminous are the Catalina Granite and Knagge pluton. Younger dikes have not as yet been dated in the study area. The plutonic rocks punctuate the tectonic evolution of the study area, and thus provide ages for some tectonic transitions.

Leatherwood Granodiorite

Mesocratic equigranular granodiorite is widespread in the study area as a large igneous sheet. The term Leatherwood quartz diorite was introduced for these rocks by Bromfield (1952), probably from the Leatherwood mines in the Marble Peak area. All related rocks in the study area are assigned to the Leatherwood, as they are similar even in petrographic details where these are preserved. Rock compositions are granodioritic in the current IUGS classification.

Distribution. The main body of Leatherwood is a large slightly discordant sheet that extends from the range crest to Geesaman Creek on the north side of the range, southeast of Alder Creek (pl. 1). This body is most easily inspected along the Control Road from the south end past the mines of Marble Peak. Toward the west this body extends to the contact with younger Catalina granite at the edge of the study area. East of this main body, porphyritic dikes and sills of Leatherwood cut Apache Group and Paleozoic rocks.

In the Catalina forerange are two Leatherwood bodies of appreciable area (pl. 1) in the Soldier Canyon and Molino basin (loc. 60) areas. A smaller unmapped body is near Sycamore Reservoir (loc. 61). These were apparently emplaced in Oracle Granite, but relations are obscured by mylonitic deformation (fig. 43). In the forerange also are numerous deformed Leatherwood dikes and sills that could not be mapped. A persistent thin zone of Leatherwood under the younger Gibbon Mountain sill suggests a precursor Leatherwood sill in that position; the larger Soldier Canyon and Molino basin Leatherwood bodies are related to this zone in structural position and perhaps in origin. In some forerange outcrops, Leatherwood dikes have become boudins that resemble xenoliths, but original textures

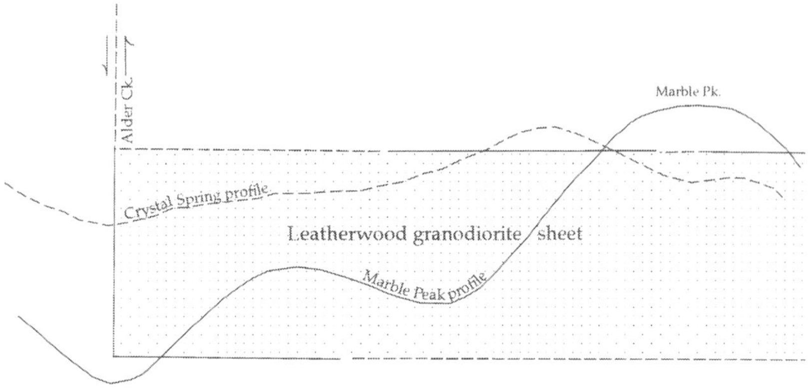

Figure 45 Diagram of the cross-sectional relations on the southeastern margin of the Leatherwood sheet, looking southwest along the sheet itself; host rocks dip out of the diagram. Two NW-SE topographic profiles (Marble Peak farther northeast, Crystal Springs Trail farther southwest) show the nature of the evidence.

Figure 44 ↑ View looking southwest from Escabrosa Limestone of Geesaman Creek area in foreground across the Leatherwood granodiorite sheet in middle distance to Marble Peak on right skyline, composed on the nearest slopes of Escabrosa and higher formations (loc. 41).

Figure 46 → Annotated photo of Marble Peak looking northwest, showing slight discordance of intrusive Leatherwood contact. Annotations as for plate 1.

are preserved and when viewed down the plunge of mylonitic lineation they have the form of disrupted discordant bodies. In road cuts just north of Molino basin, boudinaged Leatherwood dikes occur in dark granitic xenoliths (loc. 62).

In the main range, the southern margin of the Leatherwood sheet is against younger aplite and pegmatite of the Wilderness suite, and so is not an original margin. The approximate original margin can be deciphered, however, from the distribution of Leatherwood versus Oracle Granite xenoliths in Wilderness suite rocks. Such xenoliths locally are over a hundred meters long. Plate 1B shows that these Leatherwood xenoliths occur mostly in two main areas, both in the Lemmon Rock pegmatite-aplite map unit. The first is in upper Sabino Canyon as far downstream as Marshall Gulch; this is a southern extension of the main Leatherwood body farther up-canyon. In this area, remnants of the Leatherwood sheet dip south under Oracle Granite. The second area is on the south slope of Mt. Lemmon between Lemmon Rock (loc. 36) and Wilderness of Rock, where they are present throughout the Lemmon Rock pegmatite-aplite area, between two outcrop belts of Pioneer Formation. Thus a discrete Leatherwood sill must once have existed in this location.

Form. The main Leatherwood body is a slightly discordant sheet about 300 m (1000 ft) thick, intruded into Paleozoic rocks and the Apache Group. Along its southeastern edge, however, this sheet has an abrupt steep-sided termination in the valley of Alder Creek (loc. 63), where Leatherwood on the northwestern side of the valley faces a thick section of Proterozoic and Paleozoic supracrustal rocks on the other side across an unfaulted contact, with only thin sills of a porphyritic facies extending into the supracrustal sequence. Toward

Table 6 Chemical and modal analyses of the Leatherwood granodiorite; n.d., not determined.

A. Chemistry
(* , all iron reported as one oxide; + , mylonitic)

	1	2	3	4	5	6
SiO_2	62.7	64.8	62.1	62.8	n.d.	n.d.
Al_2O_3	16.1	16.3	16.6	15.8	15.0	15.5
Fe_2O_3	2.1	4.3	4.9	*	6.0	6.1
FeO	2.1	*	*	5.15	*	*
MgO	2.9	n.d.	n.d.	3.28	2.7	2.9
CaO	4.7	2.38	3.08	4.75	3.8	4.0
Na_2O	3.7	3.94	3.48	3.82	2.9	2.9
K_2O	2.8	2.40	2.28	3.30	1.8	1.9
TiO_2	0.60	0.61	0.72	0.66	0.7	0.7
P_2O_5	0.22	0.13	0.28	0.29	0.3	0.4

	7	8+	9+	10	11
SiO_2	n.d.	n.d.	64.4	64.41	69.35
Al_2O_3	14.7	13.0	15.2	15.95	14.93
Fe_2O_3	4.8	3.6	1.18	1.46	1.19
FeO	*	*	2.10	3.81	3.07
MgO	2.2	1.3	1.46	2.45	0.94
CaO	3.8	2.7	2.98	5.36	3.04
Na_2O	3.0	2.9	3.41	3.39	4.67
K_2O	2.1	1.6	2.89	1.45	1.48
TiO_2	0.6	0.4	n.d.	0.62	0.48
P_2O_5	0.1	0.3	0.15	0.20	0.15

B. Modes
(n.r., none reported; tr, trace)

	1	2	3	12	13
Quartz	21.0	19	16	28	22
K-feldspar	7.5	12	6	8	5
Plagioclase	47.0	51	41	31	42
An value	38	n.d.	n.d.	n.d.	34-42
Biotite	20.0	9	9	18	19
Hornblende	3.5	n.r.	n.r.	5	1
Epidote	n.r.	4	9	4	1
Sphene	tr	1	1	n.r.	<1
Apatite	tr	tr	tr	n.r.	n.r.
Opaque	2.0	1	tr	n.r.	n.r.

Descriptions

1. Sample BR2 of Banks (1980), from near Lombar Hill, by rapid rock analysis.
2. Sample AZ-LWD1 of Thomas Heidrick, reported by Keith and others (1980).
3. Sample AZ-LWD3, as in #2.
4. Sample FT GB (loc. 66), from Control Road, analyzed by XRF for this study.
5. Sample 155864, reported by Coney and Reynolds (1980), analyzed by ICP spectroscopy and recalculated to oxides; from range crest.
6. Sample 155869, as in #5, from range crest.
7. Sample 155921, as in #5, from range crest.
8. Sample 155917, as in #5, mylonitic from Sabino Basin.
9. Sample 155830 of "quartz diorite" tentatively assigned to Leatherwood by Coney and Reynolds (1980), and mapped as such in this study. Analyzed by atomic absorbtion; mylonitic, from Molino Basin.
10. Average hornblende biotite tonalite of Nockolds (1954).
11. Average biotite tonalite of Nockolds (1954).
12. Sample S-2-1 of Suemnicht (1977), from Canada del Oro.
13. Average of 26 samples reported by Hanson (1966).

the northeast this steep margin crosses a ridge (of Lombar Hill) as Alder Creek diverges to the east (fig. 41).

The approximate northeastern margin of the sheet is formed by the Geesaman fault; however, in detail the main body of Leatherwood cuts this fault and is in intrusive contact with rocks on its northern wall (loc. 64). In the Marble Peak area the top of the Leatherwood sheet remains approximately parallel to Paleozoic stratigraphy across a syncline, so that the upper contact dips rather steeply southwest.

East of the main Leatherwood body, Leatherwood dikes cut the Geesaman fault and a splay of that fault (loc. 65). The Edgar fault is also cut by Leatherwood bodies (pl. 1).

Where the exposure geometry of the Leatherwood sheet permits determination of the rocks above and below the sheet in the same area, these rocks are the same (fig. 44, pl. 1). For example, where the top of the Leatherwood cuts the Bolsa-Abrigo contact on the shoulder of Marble Peak, the base of the Leatherwood cuts the same contact in Alder Canyon below. This shows that the Leatherwood was emplaced by dilation, and that the steep-sided southeastern margin of the sheet must also be a transform fault. Movement was only above the structural level of the sheet (fig. 45), as footwall rocks are apparently not displaced (pl. 1).

The orientation of the Leatherwood sheet, except in the syncline at Marble Peak, defines a plane dipping gently northeast, but less steeply than the sedimentary rocks that it intrudes, such that the sheet cuts progressively downsection to the southwest (pl. 1), with a few irregularities (fig. 46). At its northeastern extremity the intersection of its base with the steep margin is at the structural level of the Earp Formation. At its southwestern extremity near Sykes Knob, the intersection of its top with this steep margin is in Pioneer Formation, downsection approximately 860 m (2800 ft), in spite of some pre-Leatherwood structural repetition. Outlying areas of Leatherwood continue this trend—the Leatherwood xenoliths in Lemmon Rock pegmatite-aplite at Lemmon Rock are at the level of Pioneer Formation also (loc. 36), and those near Marshall Gulch form a zone under a zone of Oracle Granite xenoliths (pl. 1B). The bodies of Leatherwood in the forerange intrude Oracle Granite. In the area north of the range crest, the main Leatherwood sheet has engulfed several faults that predated Leatherwood intrusion. These are discussed further in the structure section.

Description and composition. The Leatherwood granodiorite is an equigranular mesocratic rock with an average grain size of about 1 to 2 mm. Chemical and modal compositions in table 6 include granodiorite and lesser monzodiorite. Comparison of Leatherwood compositions with those of average quartz diorites or tonalites (Nockolds, 1954) of the older terminology suggests that the Leatherwood is more comparable to granodiorite in either the older or the current terminologies.

The Leatherwood has been described petrographically by Hanson (1966). Plagioclase is characteristically zoned. Biotite and some amphibole appear secondary. Myrmekite is common. Microcline is interstitial or replaces plagioclase. Epidote occurs both as euhedral grains and replacement mosaics. Euhedral epidote occurs as 1) separate grains, but not as phenocrysts in porphyritic Leatherwood sills, 2) inclusions in plagioclase, where its abundance is inverse to calcium content of plagioclase (Hanson, 1966), and 3) inclusions in secondary minerals such as muscovite and green biotite (cf. Anderson and others, 1988). Thus all the epidote in the Leatherwood is probably secondary.

West of Alder Creek, in and near the main intrusive sheet, Leatherwood has an equigranular appearance even in dikes and chill zones. In such zones, average grain size decreases toward intrusive contacts except for a few altered phenocrysts, apparently amphibole. At most intrusive contacts, however, the Leatherwood is not much chilled. The Leatherwood contains minor percentages of unidentified xenoliths.

Zones of calc-silicate skarn in calcareous host rocks have formed adjacent to intrusive Leatherwood contacts by contact metamorphism. Locally these contain mineral deposits.

The Leatherwood shows a strong spaced foliation that is subhorizontal to gently northeast-dipping (that is, generally parallel to the margins of the body) where it is strongest. Structural intervals of about 30 m (100 ft) may be free of this foliation, but in adjacent intervals it is strong as anastomosing zones of deformation. Dikes of Wilderness-suite pegmatite-aplite dikes cut foliation in the Leatherwood.

In the Catalina forerange, the Leatherwood is mylonitized (fig. 43, loc. 60). Even so, zoned feldspars, euhedral epidote, and other characteristic properties locally survive. Chemically, mylonitic Leatherwood seems to contain less iron, magnesium, and calcium (table 6). Probably this difference is primary, as no such difference was noted in mylonitic Oracle Granite (table 1). The mylonitic fabric is described in a following section.

Porphyritic sills. East of Alder Creek, porphyritic mesocratic sills are widespread (pl. 1). Some of these bodies are demonstrably rooted to the west in the main intrusive sheet of Leatherwood granodiorite. Individual porphyry sills are from 30 to perhaps 120 m (100 to 400 ft) thick, typically as subhorizontal to northeast-dipping bodies. These rocks typically contain phenocrysts of plagioclase, biotite, and quartz in a mesocratic but micro-aplitic groundmass. One such sill, in the Edgar and Atchley drainages, has metamorphosed host-rocks of the Abrigo Formation to coarse diopside-bearing assemblages.

Age. The age of the Leatherwood is about 68.5 Ma, based on a Rb/Sr whole-rock isochron derived from four samples from the southeastern flank of Marble Peak (Keith and others, 1980; Reynolds and others, 1987). The uncertainty of the date has been listed by others as $\pm$8 Ma (for example, Anderson and others, 1988). A similar isochron that includes samples of similar "quartz diorites" from the adjacent Tortolita Mountains gives 71.5 Ma (Keith and others, 1980; Reynolds and others, 1987). Guerin and others (1990) mention a U-Pb date on zircon of 64.2 $\pm$ 1.9 Ma for Leatherwood. However, their data (in an unpublished manuscript by Guerin and others) suggest to George Gehrels (oral. commun., 1993) that the uncertainty is actually much larger. Assuming the Cretaceous-Tertiary boundary to be at about 66 Ma, the Leatherwood seems likely to be latest Cretaceous.

Origin. If the sedimentary host rocks of the Leatherwood were approximately horizontal when the tabular body was intruded, then its original inclination would have been about 20° to the southwest. This seems likely to be the direction from which magma was supplied. Intrusion inflated the section of host rocks by about 300 m (1000 ft) in an area bounded on the southeast by Alder Creek and on the northeast by the older Geesaman fault. Composition of the Leatherwood is typical for similar rocks of this age in southeastern Arizona.

Anderson and others (1988) proposed that the emplacement depth of the Leatherwood was 21$\pm$1 km based on geobarometric considerations, including the aluminum contents of hornblende and the paleodepth implications of magmatic epidote. Palais and Peacock (1990) argued that 7.5 to 14 km is a more reasonable range, based on other geobarometric factors such as the presence of andalusite and on the thickness of available overburden. The following considerations suggest to me that emplacement was rather shallow in the type area near Marble Peak: 1) skarn-type contact metamorphism, 2) chilled contacts, and 3) association with *lateral* porphyritic facies. In addition, strata available to form overburden at the northeastern end are not very thick—the upper Naco Group is not preserved anywhere in the area, and preserved Bisbee Group is thin. The lower member of American Flag Formation sits in part structurally *under* the Paleozoic section that Leatherwood intrudes, and higher members define a coeval

Laramide ground surface. Relations of both American Flag Formation and vent facies of Rice Peak porphyry suggest this Laramide land surface was about the level of Escabrosa Limestone in the terrane north of the Geesaman fault that the Leatherwood intruded.

The cross sections (pl. 1) show the relation of the Leatherwood sheet to pre- and post-Leatherwood structures and to the formations that indicate a roughly coeval land surface. Figure 47 shows extrapolations of this surface, corrected for post-Leatherwood deformation. I conclude from this reconstruction that *some* Leatherwood was emplaced at depths as shallow as 1900 m (6300 ft). This shallow depth is consistent with above-described features of the Leatherwood itself, including evidence of secondary epidote.

Porphyries of Rice Peak

The term Rice Peak porphyry is commonly used to refer to porphyritic rock that occupies considerable areas between the Geesaman and Mogul faults (pl. 1), and which are thought by some workers to represent a shallow-level equivalent of the Leatherwood granodiorite. Keith and others (1980) introduced the term into the literature as Rice Peak granodiorite porphyry. Continuation of informal usage seems best, as the unit includes at least two rock types that may be of different ages and origins, and the unit is too poorly understood for formal introduction. The unit as mapped by previous workers includes pyroclastic rocks of the Pioneer Formation; on plate 1 an attempt is made to sort these units out.

Distribution and form. Bodies of Rice Peak porphyry are mostly restricted to the area between the Geesaman and Mogul faults. They are most common toward the south, near the Geesaman fault, for which they locally form the northern (foot-)wall. Thus the distributions of Leatherwood granodiorite and Rice Peak porphyries overlap slightly where Leatherwood has engulfed the Geesaman fault.

The Rice Peak bodies form a network of stocks and sills that are interconnected at the present ground surface via constrictions as thin as 6 m (20 ft). All the bodies, however, seem to be appendages of a single master sill or laccolith that closely follows the unconformity at the base of the Apache Group and to a lesser extent the pyroclastic facies of the upper Pioneer Formation (pl. 1, fig. 48). Two discordant bodies are found below and two above these horizons. Those below may have been the feeders. The largest and most complex bodies are those above the sill. At both map and outcrop scale, these bodies are choked with inclusions, many apparently cognate.

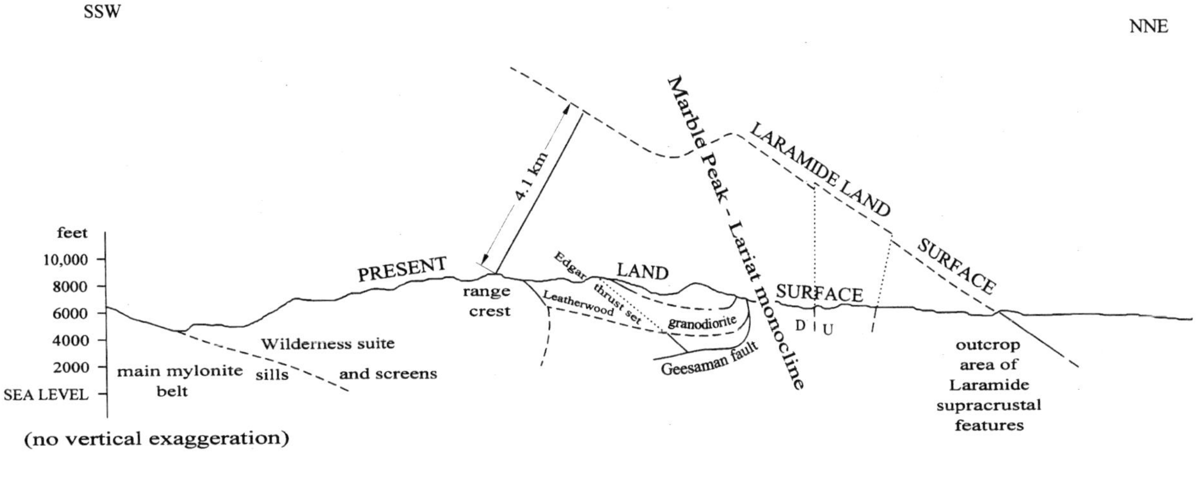

Figure 47. Simplified cross-section taken from B-B' of plate 1 showing a proposed late Laramide (following intrusion of Leatherwood granodiorite) land surface, based on the intersection of Laramide supracrustal rocks with the present land surface. The features include a vent facies of Rice Peak porphyry (older variety) and the middle member of American Flag Formation. Extrapolation of the paleo-surface to the southwest is possible because several major structures are pre-Leatherwood, and thus do not require reconstruction. The younger structures for which restoration is required are mostly minor; most important is the post-Leatherwood part of movement on the Marble Peak-Lariat monocline. Near and south of the range crest itself, post-Leatherwood Laramide fabrics complicate land-surface reconstruction. This method of Laramide restoration to the range crest is required because of extensive pre-Leatherwood faulting, other early phases of Laramide deformation, and intrusions (especially sills) of earlier Rice Peak porphyry. The Leatherwood sheet itself forms the most reliable datum southwest of the Geesaman fault because its orientation is regular and it cuts many older structures.

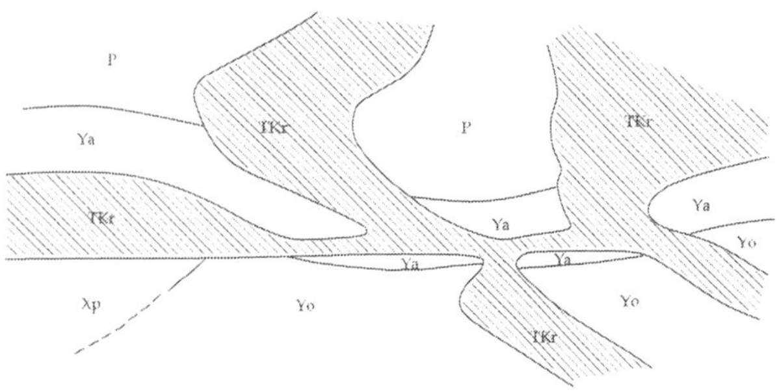

Figure 48 Diagram of Rice Peak porphyry (TKr) bodies looking northeast down the dip of host rocks, with post-Rice Peak deformation restored. Pz includes all Paleozoic; Ya all Apache Group; Yo, Oracle Granite; Xp, Pinal Schist. The larger Rice Peak bodies are rooted on the unconformity at the base of the Apache Group.

Description. The Rice Peak porphyries are typically green-gray porphyritic rocks, most commonly altered both deuterically and by dynamic metamorphism. Contained phenocrysts are plagioclase, biotite, and hornblende; Suemnicht (1977) recorded orthoclase and quartz phenocrysts locally. The groundmass may be either microlitic or aplitic. Alteration is to chlorite, epidote, calcite, and sphene. This alteration has prevented reliable modal analysis. Table 7 lists analyses of three specimens of different types.

Porphyritic Leatherwood sills contain more quartz and biotite than Rice Peak porphyries. Chemically, some Rice Peak bodies seem similar to the Leatherwood. However, compositional data are too sparse for meaningful compositional comparison.

Two varieties of Rice Peak porphyry are apparent; both are locally present within the same outcrop. In the large stock east of Rice Peak, for example, darker porphyry with abundant plagioclase and a few hornblende phenocrysts occurs as abundant inclusions in more trachytic-textured porphyry with biotite, amphibole, and plagioclase phenocrysts (loc. 68). The darker and older variety of porphyry appears to predominate in the concordant body emplaced along the base of the Apache Group, and in some other bodies.

Deformation has converted some areas of Rice Peak porphyry into phyllitic rock, locally with remarkably flattened phenocrysts (loc. 69). The foliation is mostly sub-horizontal. This fabric increases in strength and dips south near the Stratton detachment fault. Locally, the younger variety of Rice Peak porphyry appears to cut foliation in the older variety (loc. 70).

Intrusive contacts of Rice Peak porphyries into sedimentary country rocks are associated with very little contact metamorphism, even where wallrocks are calcareous, and minor hydrothermal alteration. Where Pioneer Formation argillites are the wallrock, a faint reddening due to oxidation is observed. Chill zones at the contacts are subtle, with phenocrysts persisting throughout.

In a small Rice Peak porphyry body in Escabrosa Limestone adjacent to Stratton Canyon, massive porphyry apparently of the older type grades upward (northward) into volcanic breccia with porphyry clasts in volcanic groundmass (loc. 71). Thus the Rice Peak vented locally at the approximate structural level of the Escabrosa Limestone. This has implications for the emplacement depth and Laramide erosion level for several other units (fig. 47).

Compositional variations. Table 7 includes one analysis each of the younger and older varieties of Rice Peak porphyry. The younger is more alkali-rich and poorer in iron and magnesium than a specimen of the older.

Age. No radiometric ages on the Rice Peak porphyries have been published. Unruh (this volume) attempted U-Pb dating with zircon, but could suggest only a broadly Laramide age (71 ± 8 Ma) for the older

Table 7 Chemical analyses of the Rice Peak and San Manuel porphyries; *, all iron reported as one oxide.

	Rice Peak porphyry			San Manuel porphyry		
	1	**2**	**3**	**4**	**5**	**6**
SiO_2	61.8	65.9	66.6	61.8	64.88	67.3
Al_2O_3	16.8	17.5	16.3	16.5	15.01	16.0
Fe_2O_3	3.0	*	*	3.3	1.06	*
FeO	2.6	4.70	3.98	1.2	2.42	4.15
MgO	?	2.48	1.97	2.6	1.44	3.22
CaO	5.0	2.22	1.95	3.6	2.59	1.83
Na_2O	3.8	3.95	5.67	4.3	3.02	4.16
K_2O	1.9	2.34	2.63	2.5	3.14	2.52
TiO_2	0.70	0.59	0.48	0.62	0.56	0.57
P_2O_5	0.26	0.21	0.23	0.22	0.19	0.16

Descriptions

1. Rice Peak (granodiorite) porphyry from the large body northeast of Rice Peak, from Creasey (1967).
2. Rice Peak porphyry, older variety, sample FT BH (loc. 67), sill in Pioneer Formation near Oracle Ridge, analysis by XRF for this study supplied by Daniel M. Unruh.
3. Rice Peak porphyry, younger variety, sample FT820 (loc. 52), sill in American Flag Formation middle member, analysis by XRF for this study supplied by Daniel M. Unruh.
4. San Manuel (granodiorite) porphyry from relatively fresh exposures in the Kalamazoo-Purcell area from Creasey (1965, NOT 1967).
5. San Manuel (monzonite) porphyry from drill cuttings, biotite zone structurally above the San Manuel ore body, from Schwartz (1953, p. 10 and 25) .
6. San Manuel porphyry (apparently type B) from fresh outcrop in Purcell window, sample FT 84 (loc. 72), paleomagnetic site 1 of Force and others (1995), analysis by XRF for this study supplied by Daniel M. Unruh.

variety. Even such an approximate age has implications for the ages of structural features.

The age of the younger variety of Rice Peak porphyry could not be determined radiometrically in our samples. However, a minimum age is implied by dikes of the younger variety south of the Geesaman fault, which are cut by Eocene pegmatites.

Relation to Leatherwood granodiorite. The relative ages of Rice Peak and Leatherwood are ambiguous if the Rice Peak is considered a single unit. The relation across the Geesaman fault suggests that the Rice Peak is older; dark plagioclase-amphibole porphyry locally forms the footwall of the fault, which elsewhere is intruded by Leatherwood. However, on the eastern flank of Lombar Hill, trachytic porphyry containing biotite forms dikes in Leatherwood (pl. 1). These conflicting relations are consistent with the relative ages of the two porphyry varieties east of Rice Peak, however, and thus it is likely that one Rice Peak variety is older and one younger than Leatherwood (fig. 49). The older bodies are probably Late Cretaceous, based on the approximate date by Unruh (this volume) and the relation of the older Rice Peak bodies to better-dated Leatherwood.

Relation to American Flag Formation. In the description of American Flag Formation, I noted that of three members, the two lower members are cut by porphyry, and the two upper members contain clasts of it. The porphyry that intruded along the base of the middle member (loc. 52) is a trachytic-textured porphyry, i.e. the younger variety, and the clasts in the middle and upper members appear to be from the older variety. These relations are consistent with those elsewhere. The relative chronologies of American Flag accumulation, Rice Peak intrusion, and Leatherwood intrusion are shown in figure 49.

Origin. The Rice Peak porphyries are mostly hypabyssal intrusions disposed about a master sill or laccolith at or near the base of the Apache Group (fig. 48). They appear to have vented locally. Rice Peak porphyries may be close to Leatherwood granodiorite in composition and age, but varieties appear to be either younger or older than Leatherwood.

The lack of contact metamorphism and hydrothermal alteration associated with Rice Peak porphyries may be due to a paucity of excess heat. Four factors may have minimized such impacts: 1) Most homogeneous bodies are thin, and most larger bodies are collages formed by successive intrusion of small bodies, 2) Abundant inclusions absorbed heat, 3) The portion of Rice Peak porphyry that vented would have lost most of

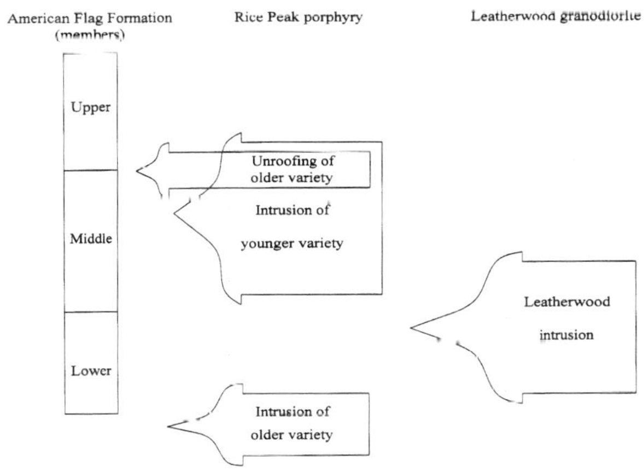

Figure 49 Correlation diagram for different phases of American Flag Formation accumulation, Rice Peak porphyry intrusion, and Leatherwood granodiorite intrusion.

its heat and volatiles, and 4) The presence of phenocrysts in chill zones suggests that phenocrysts had already crystallized at the time of emplacement, so that their latent heat of crystallization was not available at the emplacement site.

Porphyries of San Manuel

In the San Manuel mine area is a dark gray porphyry, divided into two segments by the San Manuel fault, informally known as the San Manuel porphyry. The porphyry copper deposits of the San Manuel district are centered on this porphyry. It is hydrothermally altered to varying degrees along with adjacent Oracle Granite that it intrudes.

Distribution. The San Manuel porphyry has been separated by extensional movement on the San Manuel fault into two segments known as the San Manuel segment and the Kalamazoo or Purcell segment. In both these segments, the porphyry is mostly covered by younger Cloudburst and San Manuel Formations. The San Manuel segment is intermittently exposed to the southeast in the horst between the San Manuel and Cholla faults where it locally intrudes probable American Flag Formation. Dikes of similar porphyry intrude Oracle Granite just north of the town of Oracle (pl. 1).

Form. The form of intrusive bodies of San Manuel porphyry is described by Force and Dickinson (1994) and Force and others (1995). The intrusion that is related to the San Manuel orebody, and probably the Kalamazoo segment thereof, is irregular in detail but tabular in general form, with a strike of about N. 58° E.

and a dip of about 47°to the southeast, measured on the present base of the body. The top of this body is poorly known, but appears to be compound, such that the porphyry dike that is related to copper ore is about 300 m (1000 ft) thick.

The intrusion that crops out in the Kalamazoo or Purcell area apparently strikes about N. 45° to 50° W., and dips steeply northeast. This body probably predates copper mineralization and lies above the ore-related porphyry.

Description. The San Manuel porphyry is variously reported as granodioritic, monzonitic, and quartz monzonitic in composition. Plagioclase, commonly zoned, is the most abundant phenocryst. Biotite, chloritized even outside the area of hydrothermal alteration, is commonly in euhedral books. Amphibole is locally present but commonly altered. The groundmass consists of quartz, K-feldspar, plagioclase, and mafic minerals, in textures varying from microlitic to aplitic. Hydrothermal alteration is extensive and varies from chlorite-carbonate-epidote through quartz-sericite to biotite-K-feldspar assemblages (Lowell and Guilbert, 1970). Only propylitic assemblages are common at the present levels of erosion.

The porphyritic and variably altered nature of San Manuel porphyry has prevented accurate modal analysis. Table 7 shows analyses of three relatively fresh samples. Compositional variation is considerable, probably as a function of discrete intrusions and different alteration intensities.

In detail, two rock types comprise the unit (Thomas, 1966; Lowell and Guilbert, 1970). The first, called type A by Lowell and Guilbert, contains more and larger zoned plagioclase phenocrysts in a microaplitic groundmass. The second, type B, has a microlitic groundmass and unzoned plagioclase. Both types form roughly tabular bodies. Type A porphyry in the San Manuel mine dips southeast, whereas type B porphyry exposed in the Purcell window (loc. 72) and in the south fork of Mammoth Wash (loc. 55) dips northeast. Type B is mineralized but geometrically little related to ore, and thus is probably older. Cross-cutting relations between two types of porphyry dikes north of Oracle also suggests the older age of type B.

Type B porphyry intrudes but is roughly concordant to the northwest-striking unconformity between Oracle Granite and overlying (probable) American Flag Formation. In contrast, apparently younger type A porphyry is discordant to this unconformity, and must cut it near the Cholla fault (pl. 1).

Tight contacts of (type B) San Manuel porphyry with Oracle Granite can be observed in a number of places in the Kalamazoo or Purcell area. A chill zone up to a meter thick may either contain few phenocrysts or contain phenocrysts that become finer grained toward the contact. The texture of San Manuel porphyries and their chill zones suggest that they were intruded as largely aphyric magmas, followed by phenocryst growth in place.

Age. Previously published ages on the San Manuel porphyry (Reynolds and others, 1986; corrected by S. C. Creasey, written commun., 1991) were 65.4 and 69.0 Ma, from K-Ar on biotite. Unruh (this volume) has obtained zircon dates of 68.1±3.0 Ma in the Purcell window area (loc. 42; paleomagnetic site 1 of Force and others, 1995) and 67.8±2.3 Ma in the south fork of Mammoth Wash (loc. 55). Both dates are on porphyry bodies with the appearance and geometric attitude of type B.

Origin. The relation of San Manuel porphyries to Leatherwood granodiorite or Rice Peak porphryries is unclear. If the correlation of American Flag Formation in the Mammoth Wash area is correct, then the younger Rice Peak porphyry and the older (type B) San Manuel porphyry both postdate most American Flag deposition, and older Rice Peak porphyry probably predates San Manuel porphyry. Unruh's approximate radiometric date on older Rice Peak porphyry (71±8 Ma) is consistent with this conclusion.

The depth of emplacement is constrained by intrusion into probable American Flag Formation. Emplacement depth at the Oracle-American Flag unconformity would equal the total thickness of the probable American Flag. Not much thickness is exposed near San Manuel, however. The greatest thickness of American Flag preserved anywhere in the study area is 635 m, which may serve as a minimum depth of emplacement. Where San Manuel porphyry intrudes Precambrian rocks, its emplacement depth would be

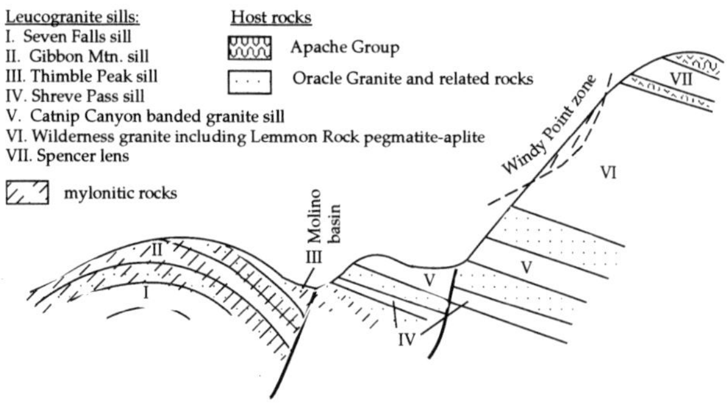

Figure 50 Configuration and stratigraphic position of the seven sills of the Wilderness suite, showing mylonitic areas. The cross-sectional relations are diagrammatic.

correspondingly greater than 635 m. The paleodepth of most type A porphyry must also be increased by the thickness of older type B porphyry intruded along the unconformity. The dip of probable American Flag Formation in Mammoth Wash can be used to reconstruct some paleo-depths; they would have been about 1700 m, for example, for type A porphyry at depth under the San Manuel fault in the San Manuel segment. The greatest paleodepths of the system would be at the lower west end of the Kalamazoo segment.

Leucocratic Granites of the Wilderness Suite

The south-facing side of the main range (fig. 4), including the forerange, is composed in large part of leucocratic granite sills of Eocene age. Seven of these sills proved to be mappable (fig. 50). One, the thickest at 2000 m (6500 ft), is emplaced approximately along the Oracle Granite-Pioneer Formation unconformity; five, for a total of 1550 m (5000 ft), are in Oracle Granite below this unconformity, and one up to 400 m (1300 ft) thick above it. The thick sill along the unconformity has a thick upper zone of pegmatite and aplite. This body is known as the Wilderness granite (of Shakel, 1978) for its occurrence at Wilderness of Rock south of Mt. Lemmon. The informal term Wilderness suite was introduced by Keith and others (1980) for all the approximately coeval leucogranite sills of the Catalinas and adjacent ranges, including the Wilderness granite.

In the forerange and parts of the main range, Wilderness suite granites and their host rocks were mylonitized in the mid-Tertiary. The mylonitic fabric is described separately, but constitutes an important barrier to understanding of these rocks that historically was quite effective until about 1980.

Distribution and form. Rocks of the Wilderness suite pervade the Catalina forerange and south-facing parts of the main range, to the extent that few outcrops of any unit are without leucogranite or pegmatite dikelets. Wilderness suite rocks intrude all Precambrian units, Paleozoic sedimentary rocks, and plutonic rocks as young as Leatherwood granodiorite (fig. 51). They also cut Laramide foliation in such units (fig. 52), though one Wilderness unit contains a weak parallel foliation itself. Dikes of leucogranite and pegmatite of the Wilderness suite extend northward a considerable distance beyond the main mapped bodies, into Leatherwood granodiorite past Lombar Hill, and the Apache Group sequence approximately to the Edgar fault (pl. 2).

All the rock bodies comprising the Wilderness suite *in the study area* form sills except in cases described with the individual bodies. Discordances are sufficiently slight that nowhere can one mapped body be said with certainty to cut another, except the Lemmon Rock pegmatite-aplite, dikes of which apparently intrude many units at diverse structural levels. The base of the Wilderness granite sill may cut structurally downward to the west.

General composition. Keith and others (1980) and Coney and Reynolds (1980) showed that the leucogranites of the Wilderness suite are peraluminous two-mica granites. For rocks of their silica content they are low in potassium. These conclusions are most cogent for two of the seven sills, the Wilderness and Thimble Peak bodies, which bear appreciable muscovite and

Figure 51 Lemmon Rock pegmatite-aplite dikes intrusive in Leatherwood granodiorite, ski road (loc. 73). Outcrop about 8 m high.

Figure 52 Specimen showing undeformed leucogranite dikelet cutting foliated dike of porphyritic Leatherwood granodiorite, Butterfly Trail (loc. 74).

Table 8 Distinguishing characteristics in hand specimen of leucogranite sills of the Wilderness suite.

Sill	Texture	Composition
Spencer	Equigranular, weakly foliated	Biotite>muscovite
Wilderness granite	Equigranular, medium to coarse	Garnet; muscovite>biotite
Catnip Canyon banded granite	Primary banding	Biotite>>muscovite
Shreve Pass	Equigranular, medium grained	Biotite only
Thimble Peak	Porphyroclastic, coarse grained	Biotite>muscovite, garnet
Gibbon Mountain	Porphyroclastic	Biotite>>muscovite
Seven Falls	Equigranular, entirely ductile fabric	Biotite>>muscovite

garnet. Each of the seven sills has some character of its own (table 8), almost to the point that they can be identified from hand specimen. The sills are described below in ascending structural order, using names introduced by Peterson (1968) where appropriate. For the bodies between the Thimble Peak and Wilderness sills, the terminology is introduced in this report, and the mapping is entirely new.

Seven Falls sill. The lowest leucogranite sill, indeed the structurally lowest unit exposed in the Catalinas, is the Seven Falls body, exposed in the forerange in the axis of an antiform in the floors of Sabino and Bear canyons (figs. 53, 54; locs. 75, 76). An additional exposure shown by Peterson (1968) in Soldier Canyon is not the same sill. The thickness of the Seven Falls body is unknown because the base is not exposed, but is at least 150 m (500 ft) in the uplifted block between Sabino and Bear canyons. Its roof is against deformed Oracle Granite. Its appearance is somewhat distinctive in its strong biotite lineation and platy quartz but the absence of any brittle component in its fabric. Thin sections verify that feldspar, as well as quartz, was deformed in ductile manner in this rock.

Table 9 shows that the granite of this sill is unusually rich in plagioclase, and poor in muscovite, compared to the other sills. Its texture is equigranular, with grain size of 2 to 3 mm, and its distinctive fabric aids identification (table 8).

Gibbon Mountain sill. This sill, typically about 240 m (800 ft) thick in the area of Sabino and Bear canyons, may be as thick as 370 m (1200 ft) at Gibbon Mountain. It is best exposed as cliffs forming the upper slopes of both Bear and Sabino canyons in the forerange (figs. 53, 54)). Due to folding, this sill is also exposed at the mouth of Sabino Canyon (loc. 77). It was emplaced in Oracle Granite and Leatherwood granodiorite; Leatherwood is common at both the base (loc. 78) and the top, suggesting that this sill followed an older Leatherwood sill.

The Gibbon Mountain sill lacks garnet and contains far more biotite than muscovite (tables 8,9). It has been mylonitized everywhere it is exposed; rolled porphyroclasts of feldspar are common.

Figure 53 Annotated photo looking east of Seven Falls (7) and Gibbon Mountain (G) sills in Sabino Canyon.

Figure 54 Annotated photo looking southwest of trace of Romero Pass fault zone (RPf) across Bear Canyon drainage, showing Seven Falls (7) sill and cliffs of Gibbon Mountain (G) sill below Thimble Peak (T).

Table 9 Chemical and modal analyses of Wilderness suite leucogranites, listed by individual sills from the top, pegmatites excluded.

A. Chemistry

(n.d., not determined; *, all iron as one oxide; +, analysed by ICP, remainder by atomic absorbtion)

	Spencer granite	Wilderness granite sill								inclusion zone				Catnip Canyon banded granite		
	1	**2**	**3**	**4**	**5+**	**6+**	**7**	**8**	**9**	**+10**	**11**	**12+**	**13+**	**14+**	**15+**	**16**
SiO_2	72.1	72.9	75.2	76.4	n.d.	n.d.	74.0	76.4	n.d	74.1	75.2	n.d.	n.d.	n.d.	n.d.	74.4
Al_2O_3	15.4	14.5	14.4	14.7	15.1	12.8	14.9	14.6	13.1	14.0	15.1	13.8	15.5	15.1	14.0	15.4
Fe_2O_3	0.81	0.51	0.46	0.25	2.1	1.7	0.48	0.43	0.9	0.52	0.48	1.4	1.0	3.8	2.6	0.75
FeO	0.75	0.40	0.25	0.24	*	*	0.35	0.32	*	0.21	0.37	*	*	*	*	0.51
MgO	0.41	0.19	0.14	0.08	0.3	0.2	0.15	0.12	0.1	0.12	0.18	0.1	0.2	0.8	0.5	0.23
CaO	1.46	1.23	1.05	0.96	1.3	1.3	1.21	1.15	1.1	1.10	1.59	1.4	2.2	1.9	1.6	1.50
Na_2O	4.36	4.12	4.19	3.33	3.6	2.9	3.86	3.70	3.3	3.45	4.39	2.7	4.4	3.1	3.1	3.54
K_2O	3.39	3.42	3.63	4.10	2.8	2.4	3.54	3.64	2.8	3.47	2.86	3.3	2.3	2.8	3.0	3.92
P_2O_5	0.08	0.04	0.03	0.04	0.04	0.04	0.03	0.03	0.02	0.03	0.03	0.04	0.02	0.2	0.2	0.05
TiO_2	n.d.	n.d.	n.d.	n.d.	0.2	0.1	n.d.	n.d.	0.1	n.d.	n.d.	0.1	0.1	0.4	0.3	n.d.

B. Modes

(n.r., none reported)

	17	**18**	**19**	**20**	**21**
quartz	28	31	32.9	29.9	29.1
plagioclase	29	34.5	33.6	35.5	41.3
An content	20-25	15-25	20-25	20-25	25-30
K-feldspar	27	24.5	26.6	26.0	21.8
biotite	4	4.5	4.1	7.7	7.2
muscovite	7	4	2.4	0.3	0.1
garnet	1-2	<1	n.r.	n.r.	n.r.
opaques	1	<1	0.2	0.4	0.3

Descriptions

All chemical analyses from Coney and Reynolds (1980); samples assigned to sills for this report based on reported collection locality.

1. Sample 155856, from near Palisades ranger station, .
2. Sample 155854, near heliport on Sollers Road.
3. Sample 155852, near Rose Canyon camp.
4. Sample 155845, foliated, near Windy Point.
5. Sample 155850, foliated, near Windy Point.
6. Sample155846, foliated, between Windy Point and General Hitchcock camp.
7. Sample 155847, foliated, between Windy Point and General Hitchcock camp.
8. Sample 155848, near General Hitchcock camp.
9. Sample 155919, near base of Wilderness sill in Bear Canyon.
10. Sample 155839, inclusion-filled granite, Bear Canyon.
11. Sample 155840, inclusion-filled granite, Bear Canyon.
12. Sample 155833, inclusion-filled granite, Molino Basin.
13. Sample 155918, inclusion-filled granite, Molino Basin.
14. Sample 155836, banded granite,
15. Sample 155837, banded granite,
16. Sample 155838, banded granite, Seven Cataracts area.

Modally analysed collections

17. Average of 10 analyses of Wilderness sill, from Pilkington (1962).
18. Average of 4 analyses of Wilderness sill near Windy Point, from Banks (1980, table 1, nos. 23-26).
19. Average of 16 analyses of Thimble Peak sill, from Sherwonit (1974).
20. Average of 7 analyses of Gibbon Mountain sill, from Sherwonit (1974).
21. Average of 7 analyses of Seven Falls sill, from Sherwonit (1974).

Thimble Peak sill. The structurally highest sill restricted to the forerange, the Thimble Peak sill is named from a distinctive peak supported by cliffs (fig. 54) that forms an outlier of the main body. It is emplaced in Oracle Granite both above (fig. 55) and below. Its apparent thickness in the block between Bear and Sabino canyons is about 800 m (2600 ft); this figure is approximate because folding complicates the measurement (cross section A-A', pl. 1). Toward the western edge of the map area, this sill seems to have a bulbous termination (pl. 1).

The Thimble Peak body contains slightly more biotite than muscovite (table 9), and locally bears garnet. Coarse-grained, porphyroclastic granite is most common (loc. 80). An associated less-deformed equigranular medium-grained granite is probably younger.

Shreve Pass sill. Just north of the main splay of the Romero Pass fault zone is a relatively thin sill in Oracle Granite and lesser Leatherwood granodiorite that forms the slope north of Shreve Pass. (New geographic names have been approved by the Arizona and U. S. boards of geographic names.) The best exposure is at locality 81. The thickness is about 60 m (200 ft). The texture is typically equigranular and medium-grained, though flattened by deformation. It contains quartz, two feldspars, biotite, and minor opaques and apatite; it apparently lacks muscovite and garnet.

This sill is separated by the Romero Pass fault from all the above-described sills, so its structural position relative to them is uncertain. However, it probably overlies them, because movement on the Romero Pass fault zone is south-side-up. The thickness and character of the Shreve Pass sill do not encourage correlation with any sill of the forerange.

Catnip Canyon banded granite sill. A leucogranite sill characterized by primary banding structurally overlies the Shreve Pass sill but is separated from it by Oracle Granite, and is directly overlain by the inclusion zone described in the next section. This sill is named from the canyon where it is best exposed on the Catalina Highway (loc. 82). Its apparent thickness is about 300 m (950 ft) in the Seven Cataracts region of Bear Canyon, a minimum figure limited by the northern splay of the Romero Pass fault zone.

Banding is formed by an alternation or grading of layers richer and poorer in biotite, and locally of feldspar grain size. Alternation is typically on a scale of 2 to 50 cm (1 to 20 in). The banding seems to be an original feature of the rock; later pegmatite bands are present but rather sparse, and tectonic disruption is common but local. Low-angle unconformities (fig. 56) are common and are everywhere consistent with upward accumulation (i.e. lower layers truncated). Some observers think all these truncations are tectonic, however.

Both light and dark colored bands consist of quartz, finely perthitic K-feldspar, plagioclase, biotite, and minor magnetite in varying proportions. Muscovite is locally present; garnet was not seen. Chemical composition is shown in table 9. The texture is protoclastic but the strain is insufficient to cause banding (fig. 57). Indeed, one can locally still see original igneous textures.

The banding may be cumulate in origin, as cumulate banding is known from rocks of this composition (Sisson and others, 1988). If so, and if the apparent unconformities are original features, this sill accumulated upward from the base. Further study could

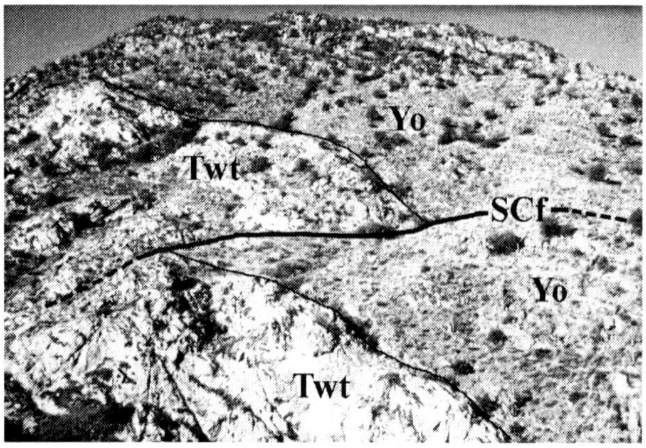

Figure 55 Upper contact of Thimble Peak sill cut by Sabino Canyon fault, Sabino Canyon looking northwest at locality 79. Twt, Thimble Peak sill; Yo, mylonitic Oracle Granite; SCf, Sabino Canyon fault.

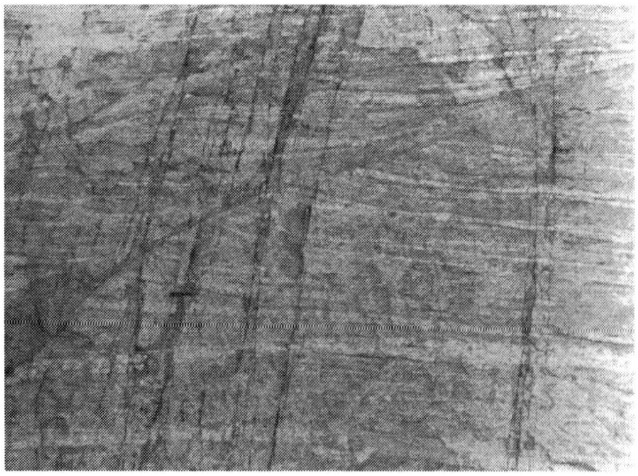

Figure 56 Banding with apparent unconformity in Catnip Canyon banded granite, north of locality 82.

Figure 57 Specimen showing texture of Catnip Canyon banded granite at locality 82.

Figure 58 Metadiabase xenolith in inclusion zone on Catalina highway (loc. 83). Xenolith (about 6 m high) contains fabrics cut by leucogranite.

Figure 59 Top of Catnip Canyon banded granite on Catalina Highway near locality 83; above it is a foliated xenolith of Oracle Granite in the inclusion zone. Field of view about 10 x 15 m.

possibly establish magma history and delineate its source.

Monazite from this sill was dated by U-Pb at locality 82 by George Gehrels (University of Arizona, written commun., 1992); the preliminary age obtained was about 46 Ma, or middle Eocene. However, subsequent work by Gehrels, M. Harrison, and R. Parrish suggests an age of 52±6 Ma based on Th-Pb on monazite, and U-Pb on zircon. An unpublished manuscript by G. Guerin and others mentions a U-Pb zircon date on leucogranite of 49.0 ± 2.6 Ma from a locality that I interpret as Catnip Canyon sill or possibly Lemmon Rock pegmatite-aplite.

Inclusion zone. Between the Catnip Canyon banded granite sill and the Wilderness granite sill is a zone composed predominantly of leucogranite but characteristically containing about 10 to 20 percent xenoliths, mostly of Oracle Granite but also of Leatherwood granodiorite, Proterozoic metadiabase (fig. 58), and Apache Group (?) micaceous fine quartzite. Laramide fabrics are present in some xenoliths. The zone is not a true plutonic unit but is described here because its distinctive character enables it to be mapped along the range front for about 12.5 km of outcrop length.

The zone is typically 120 to 250 m (400 to 800 ft) thick where it is best exposed, in the Seven Cataracts region of Bear Canyon (loc. 83) and in slopes north of Molino basin (loc. 62). Xenolith assemblages seem to be spatially homogeneous except that probable Apache Group xenoliths are toward the top in Bear Canyon. The leucogranites comprising the bulk of the zone are probably a mixture of intrusions from underlying and overlying units—both banded biotite granite and garnet-muscovite granite were seen in the zone. The relative age is unclear, but the Catnip Canyon banded granite does seem to cut most of the pegmatitic components of the inclusion zone at the contact (fig. 59). Low iron content in four chemical analyses of leucogranite from the zone (table 9) suggest that most of the analyzed specimens were Wilderness granite rather than Catnip Canyon banded granite.

Toward the western end of the zone's outcrop belt, in Sabino basin, the proportion of xenoliths relative to leucogranite host increases to about 40 percent. Further west, the zone is cut out by the Romero Pass fault.

Wilderness granite sill. This sill, slightly over 2000 m (6500 ft) thick if the Lemmon Rock pegmatite-aplite is included, forms the steep escarpments of the south face of the main range (fig. 4), and gives its name to the entire suite of leucogranite sills in this and adjacent ranges. It is named from Wilderness of Rock south of Mt. Lemmon, where it crops out in profusion. A more accessible little-deformed outcrop is at locality 84.

The body is fairly homogeneous, except for pegmatitic facies; its greatest variations are those of deformation. A rusty-weathering zone is present at the base of the sill. Trends in composition reported by Coney and Reynolds (1980) result from their having included the Catnip Canyon banded granite sill through the Wilderness sill as one unit.

The Wilderness sill is the only one that can be correlated across the Romero Pass fault zone; just beyond the western edge of the study area it is present in both the main range and the forerange across the fault (fig. 60). Thus the Wilderness sill serves as a structural datum for all the others. In the forerange portion, however, the Catnip Canyon banded granite body is apparently missing.

Typically the Wilderness granite is medium- to coarse-grained and equigranular. The main constituents are quartz, plagioclase, and non-perthitic microcline. Characteristically it contains as much or more muscovite as biotite (table 9). Most of this muscovite seems to be primary (but see Banks, 1980). In most areas, Wilderness granite contains about 1 to 2 percent garnet that is euhedral and 0.1 to 2 mm in grain size. Chemically the granite is peraluminous. Magnesium and iron contents are quite low, and distinct from those of the Catnip Canyon banded granite (table 9). Other aspects of the chemical character of the Wilderness granite sill are described in some detail by Coney and Reynolds (1980).

The age of the Wilderness granite appears to be about 47 ± 4 Ma, based on a U-Pb monazite age of 44.6 ± 1.5 Ma (Shakel and others, 1977; as reported by Reynolds and others, 1986) at locality 84, a zircon age of very approximately 50 Ma (interpreted by Keith and others, 1980, from data of Catanzaro and Kulp, 1964), and ages listed below for the related Lemmon Rock pegmatite-aplite.

Anderson and others (1988) suggested an emplacement depth of about 15 km for the Wilderness sill based on garnet composition. The location of their sample within the 2 km thick sill is not given. Palais and Peacock (1990) suggested less than 14 km based on aureole metamorphic assemblages at the top of the sill. Figure 47 suggests a Laramide depth of about 4.1 km for the structural level later intruded by the Wilderness sill.

Lemmon Rock pegmatite-aplite. Along the top of the Wilderness granite sill are pegmatites and banded aplites that are probably closely related to Wilderness granite in origin. Shakel (1978) called these rocks Lemmon Rock leucogranite; my informal name reflects a different emphasis. They form a coherent map unit in two areas. The first forms the upper part of the southern

Figure 60 Annotated photo looking south showing Wilderness granite (Tww) across the Romero Pass fault zone (RPf), partially hidden), west fork Sabino Creek. Yo, Oracle Granite.

Figure 61 Leatherwood granodiorite inclusions in Mt. Lemmon body of Lemmon Rock pegmatite-aplite, (loc. 36). A. Cliff face, B. Detail showing progressive digestion. Note skeletal biotite crystals in pegmatite.

face of Mt. Lemmon, below Lemmon Rock (loc. 36). The second is in upper Sabino Canyon below Summerhaven (locs. 85–87). The two areas are probably continuous under supracrustal rocks in the Carter Canyon area. Rocks in the two areas are similar but detailed structural relations differ.

In the Mt. Lemmon body, pegmatite and aplite were emplaced as a sill approximately along an unconformity between Pioneer Formation and overlying Bolsa Quartzite. Pioneer Formation forms a screen at the base of the unit that separates it from Wilderness granite. This sill has an outcrop length of about 3 km and an apparent thickness of 240 to 300 m (800 to 1000 ft). It cuts structurally upward to the east before disappearing under Paleozoic rocks. Many large xenoliths of Leatherwood granodiorite, elongate north-south, occur within this body (fig. 61), showing that the Lemmon Rock sill followed an older Leatherwood intrusion. Young (1988) described this body in some detail.

The Sabino Canyon body, with approximately the same thickness, extends along strike about 6 km, fingering out into Wilderness granite toward the east and dipping under supracrustal rocks toward the north. Its structural base is a northeast-dipping contact with Wilderness granite; age relations are unclear. Its top is against Apache Group rocks, quite commonly the Scanlan Conglomerate Member of Pioneer Formation. Toward the north its top is a digitate zone of dikes in Leatherwood granodiorite. That is, the top of the Lemmon Rock unit, which is approximately concordant to supracrustal units, intersects the older Leatherwood sill, which is slightly discordant and cuts downsection to the southwest (pl. 1). Toward the north, inclusions are Leatherwood, whereas farther south they are predominantly Oracle Granite (pl. 1B). At locality 6, an intact Oracle-Scanlan contact is preserved on the roof of this body of Lemmon Rock (fig. 16B). The inclusions are seldom rotated, and their geology can be mapped in this region (pl. 1B) as if the more abundant Lemmon Rock unit were not present.

The Lemmon Rock pegmatite-aplite as mapped in the Sabino Canyon body consists of three or more generations of steep dikes typically 1 to 5 m thick, and underlying relatively massive pegmatite with subhorizontal inclusion trains (fig. 62). Aplite with garnet bands typically is found in the dikes, mostly those of intermediate age (fig. 63). The oldest generation is locally foliated (loc. 85) and is excluded from the Lemmon Rock unit.

The mineralogy of Lemmon Rock pegmatite is simple. It consists of quartz, K-feldspar, plagioclase, muscovite, and lesser biotite, with Fe-Mn garnet the only accessory mineral in most outcrops. Apatite (rimmed by garnet) was seen at two localities. Tourmaline seems to be absent, along with the economic accessory minerals present in some other pegmatites. Pegmatite mineralogy is described in greater detail by Young (1988) and Matter (1969).

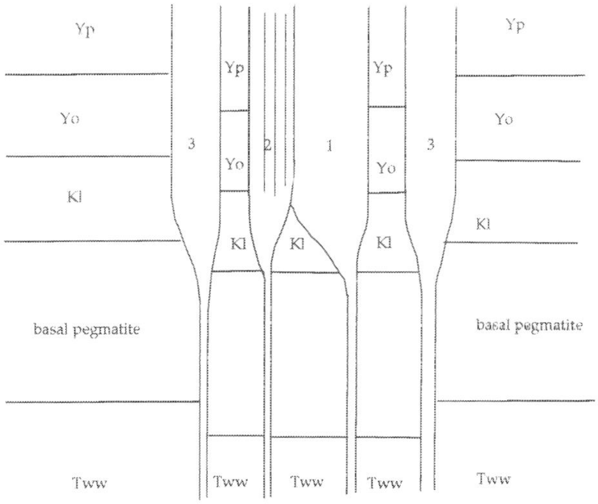

Figure 62 Diagram of cross-sectional relations of dikes at the top of the Lemmon Rock pegmatite-aplite. Symbols as for plate 1. Basal pegmatite is cut by pegmatite dike types 1 (oldest), 2 (banded aplite), and 3 (locally porphyritic), which are about parallel in dip but not in strike.

Banded aplite is common in the Lemmon Rock unit, forming as much as half of some outcrop areas. Banding may be marked by grain size or muscovite abundance differences, but commonly results from trains of fine garnet (fig. 64). Accumulation patterns are reflected by the banding of some aplites; garnet bands form stromatolite-like forms and(or) are draped over comb-textured megacrysts (fig. 64). Dikes as thin as 0.2 m may contain garnet aplite (fig. 65). Temporally, banded aplite seems to follow the formation of the oldest, locally foliated pegmatites but to precede other pegmatite (fig. 63). Spatially, aplite banding in steep dikes is at a high angle to the attitude of nearby Abrigo Formation, which is nearly horizontal. Thus the aplites have not been tilted very much and their formation is probably not a function of footwall- vs. hanging-wall. In the few pegmatite dikes that dip north, aplite is more abundant toward the base.

Pegmatite in many outcrops of the Lemmon Rock unit has a megacrystic texture in which large K-feldspar euhedra or broken euhedra, locally zoned and(or) skeletal and containing graphic quartz intergrowths, have a mica-garnet selvage. They float in a matrix of muscovite-garnet granite (fig. 66), locally aplitic and commonly banded. This texture seems to be most common in the steep dikes, especially in the Sabino Canyon body. The presence of banding in the groundmass suggests that the texture may result from disrup-

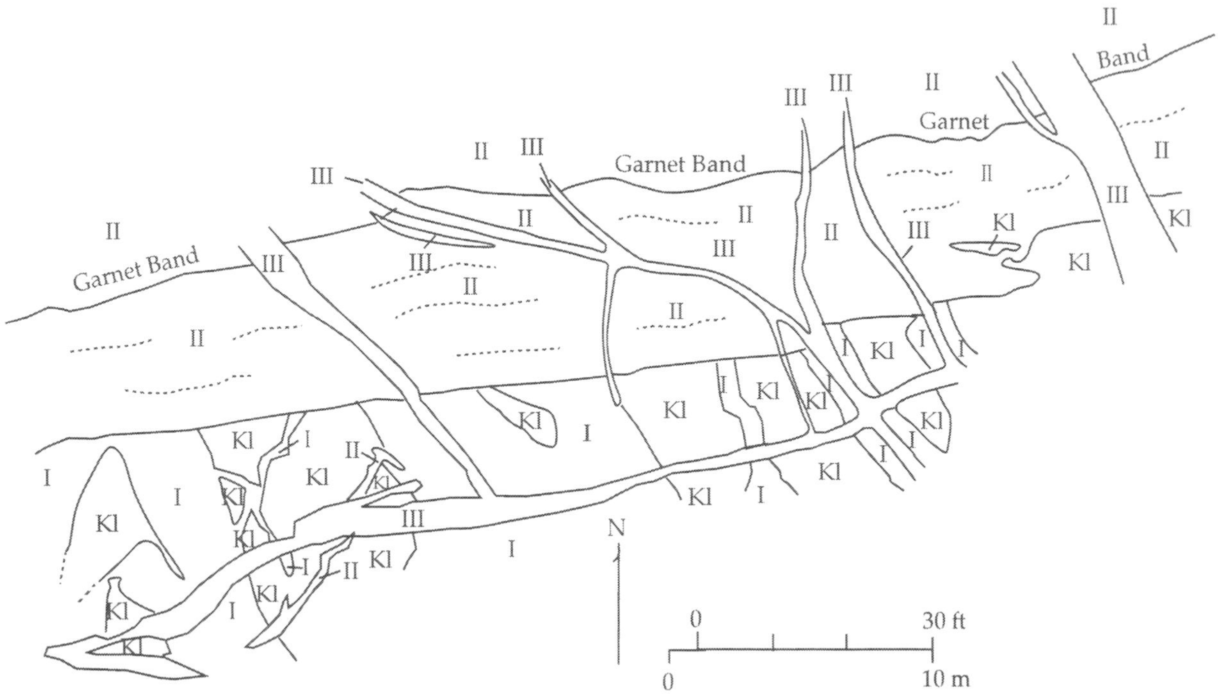

Figure 63 Map of an outcrop at locality 85 containing banded garnet aplite, showing relation to other pegmatite phases and unrotated screens and inclusions of Leatherwood granodiorite (Kl), modified from a figure in Peirce (1958). Pegmatite dike set 1 is oldest, III is youngest; II consists mostly of banded aplite.

tion of coeval aplite and comb-textured crystals as in figure 64b.

For the Sabino Canyon body a strong case can be made for a close relation of Lemmon Rock pegmatite-aplite to Wilderness granite, as pegmatite forms a thick cap to the granite. For the Mt. Lemmon body, which has the form of a separate sill, the relation is unclear. However, petrographic similarity of the two pegmatite-aplite bodies suggests that both should be considered the pegmatitic facies of Wilderness granite. Muscovite-garnet granite probably of the Wilderness granite is a subordinate component of both bodies. The two bodies give slightly different Rb-Sr ages, but this is ascribed by Keith and others (1980) to differing host rocks.

The age of the Lemmon Rock pegmatite-aplite is about 47 to 49 Ma, that is, about the same as Wilderness granite. This age is based on 1) a 47.9 $\pm$ 2.2 Ma K-Ar date on coarse muscovite (Damon and others, 1963), and 2) a Rb-Sr isochron suggesting ages of 47 to 51 Ma (Keith and others, 1980, fig. 18). Samples of Lemmon Rock pegmatite-aplite in their figure 18 define an isochron of 51 Ma or greater.

Other pegmatites. Dikes and sills of pegmatite and lesser aplite pervade much of the range, and are roughly coextensive with those of leucogranite (pl. 1B). Some bodies can be related to the Lemmon Rock pegmatite-aplite unit: 1) those within the Wilderness

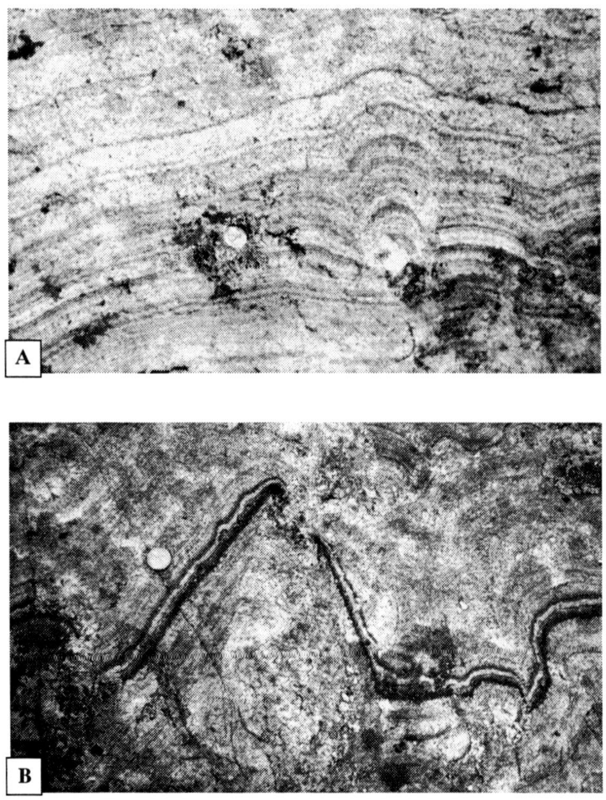

Figure 64 Banding in aplite (loc. 86), marked by garnet, showing A. Accumulation patterns; B. Banding draped over comb-textured graphic feldspar comb-megacrysts in underlying pegmatite band.

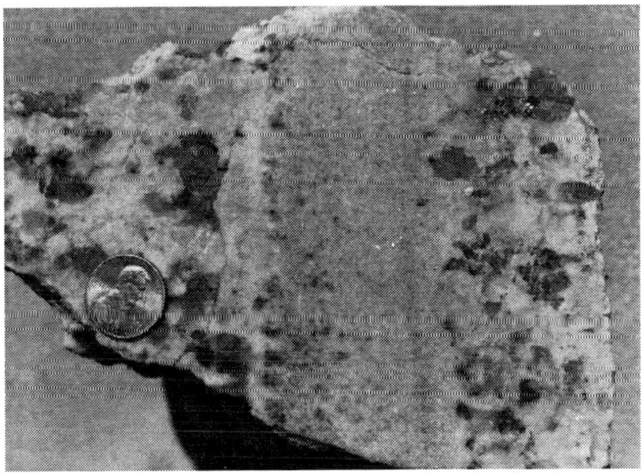

Figure 65 Banded aplite containing garnet (to right) in zoned pegmatite dike 0.2 m thick (entire thickness shown), in Leatherwood granodiorite (not shown), Loma Linda area southeast of locality 73.

Figure 66 Porphyritic texture of Lemmon Rock pegmatite, Sabino Canyon below Marshall Gulch (loc. 87). Feldspar megacrysts with graphic quartz intergrowths have garnet-mica selvages, and are set in a groundmass of muscovite-garnet granite. Field of view about 2 x 3 m.

granite sill, 2) those within the underlying inclusion zone, which is known to contain Wilderness granite, and 3) unfoliated pegmatites structurally above the Lemmon Rock unit but continuous with dike trends within the unit. The latter group includes the conspicuous pegmatite dikes along the range crest, intrusive into Leatherwood granodiorite (fig. 51) and supracrustal rocks at least as young as the Mt. Lemmon unit (Devonian?). Such pegmatites were studied by Matter (1969). Some of these upper pegmatites show the megacrystic textures characteristic of the Lemmon Rock unit, and others are zoned and include garnet aplite (fig. 65). Their roughly concentric distribution around the Lemmon Rock unit (pl. 1) itself suggests a relation.

Pegmatites and aplites in the structurally lower parts of the range may be related to other leucogranite sills of the Wilderness suite, or to younger intrusives. Those bearing garnet may be derived either from the Wilderness or Thimble Peak sills. Those that are less mylonitic than other pegmatites (fig. 67, locs. 75, 80), or some along the crest of the Forerange anticline that are ptygmatic with no consistent sense, could be mid-Tertiary, perhaps related to the Catalina or Knagge granites described below.

Where pegmatite dike arrays are undeformed, they form stockworks most commonly with one attitude predominant. In the Catalina forerange, where pegmatite and host rocks are together deformed mylonitically, the arrays are commonly deformed into apparent parallelism. However, when viewed down the mylonitic lineation, the original orientation of pegmatite bodies can still be envisioned. Probably about 60 percent of them formed sills parallel with the large leucogranite sills, and the remainder were dikes (fig. 68).

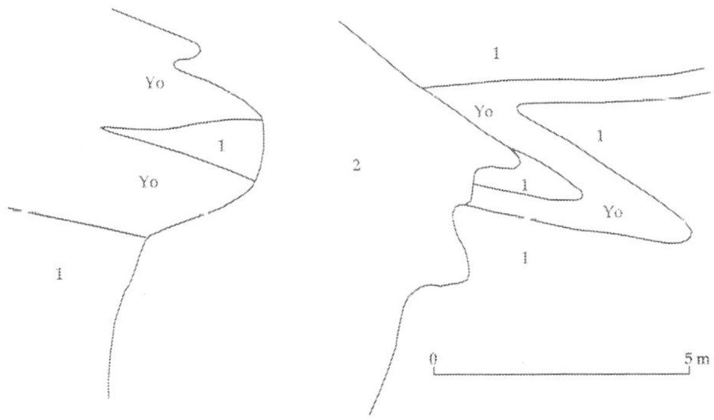

Figure 67 Diagram of outcrop relation of two pegmatites in mylonitic Oracle Granite, showing greater deformation with age of pegmatite, Catalina forerange (loc. 88). View looking down-lineation. Both pegmatites contain garnet and muscovite. Yo, Oracle Granite; 1, first generation of pegmatite, foliated; 2, second generation.

Figure 68 Down-lineation view of pegmatite dikes and sills in Oracle Granite, both mylonitized, Molino Canyon (loc. 89).

Spencer composite sill. The structurally highest member of the sills of the Wilderness suite consists of two connected lenses (pl. 1), emplaced on opposite sides of a septum of Pioneer Formation. The name was introduced by Keith and others (1980, p. 261) for Spencer Canyon camp. Discrimination as a separate body was first implied by Peirce (1958).

The Spencer granite can be delineated as a medium-grained biotite granite, with plagioclase greater than K-feldspar. Muscovite is subordinate or minor and garnet is absent. Magnetite, sphene, and apatite are common accessories.

Spencer granite is also defined by the presence of a weak fabric of probable tectonic origin, which consists of a steep east-west foliation and shallowly plunging lineation. However, the granite cuts much stronger but subparallel Laramide foliation in its host rocks. Near Mt. Bigelow, foliated Spencer granite contains rotated xenoliths of foliated Pioneer Formation.

The upper lens approximately follows the contact of Pioneer Formation and Dripping Spring Quartzite. It is about 3 km long, and apparently between 150 and 400 m (400 to 1300 ft) thick in its thickest portion. Thus its aspect ratio, about 10:1, is greater than a map view (pl. 1) implies. The lower lens, probably of similar dimensions, is limited on the south by an intrusive contact with younger Wilderness granite and pegmatite. Both lenses seem internally homogeneous. Xenoliths are uncommon in the interiors of the Spencer granite bodies, and include Leatherwood granodiorite (pl. 1B), which is not in contact with Spencer granite at the current land surface.

Relations of the upper lens to Lemmon Rock pegmatite-aplite across the screen of Pioneer Formation separating the two units were observed west of the Box Camp trail south of Spencer Peak (loc. 90). There dikes of Spencer granite cut down through the screen but are themselves cut by pegmatites of the Lemmon Rock unit. Spencer granite above the screen is cut by dikes of both Lemmon Rock pegmatite and Wilderness leucogranite.

The southern margin of the lower lens is an intrusive contact with Wilderness granite and pegmatite in which numerous Spencer granite xenoliths decrease in number to the south. Wilderness granite cuts the foliation of Spencer granite xenoliths. There is thus abundant evidence of the older age of Spencer granite than Wilderness granite and Lemmon Rock pegmatite-aplite.

Spencer granite may be related at least in age to the oldest pegmatites found in the range-crest area (i.e. type 1 of figure 62), as both are weakly foliated. Compositional correlations were not investigated.

Monazite from the Spencer granite body at locality 91 was dated using U-Pb by George Gehrels (University of Arizona, written commun., 1992); the preliminary age obtained was about 55 Ma, or early Eocene. Thus the Spencer granite is the oldest dated so far in the Wilderness suite.

Mutual age and compositional relations. No clear relation of composition to structural position of the sills of the Wilderness suite is apparent (tables 8, 9). Biotite-dominant sills are more numerous, but garnet granites with abundant muscovite predominate based on total thickness. The sills together represent about 4 km of structural thickness.

Similarly, age data so far indicate no clear trend between age and structural position, except that the highest sill is the oldest. A trend would have interesting implications for the emplacement depths, thermal history, and possibly the emplacement mechanism of the sills. Further radiometric dates should be acquired to determine a sequence and to permit thermal modeling.

Knagge Granite (informal)

Much of the eastern escarpment of the range crest is formed by a leucogranite that is apparently younger than the adjacent and similar Wilderness granite sill. It is here called the Knagge granite pluton, as it is well exposed in the Knagge trail near a mine of the same name (loc. 92). It is little foliated, and truncates most of the mylonitic fabric of the Windy Point zone in the Wilderness granite (pl. 1B). Few outcrops of it contain any pegmatite dikes.

The Knagge granite contains more biotite than muscovite, and also contains garnet. It is massive in most outcrops, but a thin mylonite zone within it transects the upper part of Barnum Rock. The pluton is weakly mylonitized near Leopold Point also.

Catalina Granite (suite)

The Catalina granite (of DuBois, 1959; McCullough, 1963) is a large body that occupies a small part of the study area, because its eastern contact is near the western edge of the study area (pl. 1). Its main outcrop area is from Canada del Oro across Samaniego Ridge to the western range front, where it forms a semicircular mass with concentric structure. External ring dikes are common, and the whole body may be a ring-dike complex.

Deformation is minor in most outcrops. However, in the area of Shovel Spring (west of plate 1) it is deformed in the Windy Point mylonite zone.

The main facies of Catalina granite is coarsely porphyritic, with subequal quartz, plagioclase, and K feldspar. Biotite and lesser hornblende, opaque minerals, and sphene constitute the ferromagnesian assemblage. Phenocrysts are orthoclase and locally hornblende. Porphyry xenoliths are common in some zones. Aplitic and alaskitic rocks form border facies and internal bands; aplite dikes are also common in some areas.

A medium-grained leucogranite forms Reef of Rock (pl. 1) along part of the eastern margin of Catalina granite, where it takes the place of the fine-grained border facies. It is apparently a distinct intrusion, probably younger (Suemnicht, 1977), but is assigned to the Catalina suite of Keith and others (1980). Intruded along the Geesaman fault west of Catalina Camp are aplite and alaskite that can be traced westward into the Reef of Rock facies of the Catalina suite.

In the area of Mt. Lemmon, the upper surface of intrusive Catalina granite is gently dipping, based on the shape of the contact toward the west (west of plate 1). If so, granite must underlie Mt. Lemmon and perhaps Summerhaven at shallow depth.

East-dipping porphyry dikes emplaced in Oracle Granite parallel the eastern margin of Catalina granite (pl. 1). Their mineralogy suggests a relation to Catalina granite, and they apparently are ring-dikes. Three lithotypes are present (loc. 93); from oldest to youngest these are: 1) a light colored biotite-hornblende-K-feldspar syenitic porphyry, 2) a mesocratic quartz-biotite-green amphibole-(tourmaline)-plagioclase microgranitic porphyry, the most voluminous type, and 3) rhyolitic porphyry, locally flow-banded, that tends to follow the Oracle-Pioneer unconformity. All three are magnetic. Suemnicht (1977) thought the dark porphyry was Precambrian. Locally these rocks are foliated but most commonly are completely undeformed, and altered only to chlorite after biotite. The rhyolitic dikes may be related to the Reef of Rock facies.

Numerous radiometric dates on the Catalina suite are mid-Tertiary, mostly between 23.5 and 28 Ma (Keith and others, 1980; Reynolds and others, 1986). These dates include U-Pb on zircon, Rb-Sr isochrons, K-Ar on biotite and hornblende, and fission-track measurements on zircon and sphene. The only U-Pb date on zircon gave 26.5 $\pm$ 2 Ma.

Other Young Igneous Rocks

Rhyolite. Dikes of rhyolite and rhyolite porphyry are widespread in the northern part of the study area. Between the Geesaman and Stratton drainages, where they were mapped by Janecke (1986; not shown on plate 1), these dikes cut rocks as young as Rice Peak porphyry. Probably they are related to the Reef of Rock facies of the Catalina suite, which introduced felsite into the Geesaman fault plane.

In the Mammoth district, rhyolite intrusives and extrusives are closely related to the Mammoth vein set (Force and Cox, 1992). There rhyolites were intruded into Oracle Granite and Cloudburst Formation, and form welded tuff and other extrusive layers a few meters thick in the lower member of Cloudburst Formation. Rhyolite also forms the matrix of intrusive breccias. Locally, intrusive and extrusive rhyolites are hard to tell apart. In some intrusive rhyolites, K-feldspar has altered to quartz and muscovite. The texture may be either aplitic or porphyritic. Intrusion of rhyolite was apparently synchronous with movement of the Cloudburst detachment fault (fig. 69) and extrusion was synchronous with accumulation of Cloudburst Formation. Some chemistry of the rhyolites is given in table 5. Just northeast of the study area, similar rhyolite was dated as 23 Ma (Dickinson and Shafiqullah, 1989).

Rhyodacitic intrusives. Intrusives called rhyodacitic by Creasey (1965) crop out in the San Manuel area, where they intrude Oracle Granite and Cloudburst Formation as dikes and plugs (pl. 1). Some of the bodies shown by Creasey prove to be extrusive. Some of the bodies in Oracle Granite are vesicular, but apparently intrusive nevertheless. Fresh specimens contain phenocrysts of plagioclase, biotite, and minor magnetite in an aplitic groundmass. A chemical analysis is included in table 5.

Biotitic granitoid dike-rock. Dark fine-grained quartz-biotite-plagioclase granitoid dikes are intruded into several fault planes in the southern part of the study

Figure 69 Rhyolite dikes (Tr) in Oracle Granite (Yo, below Cloudburst detachment fault, Cf) and Cloudburst Formation (Tcbl)) showing modest offset, Tucson Wash (loc. 94).

area, notably the Sabino, Bear, and Romero Pass post-mylonite faults. Commonly the dikes contain epidote-group minerals, magnetite, sphene, tourmaline, and (or) fluorite. Locally the dikes are brecciated or foliated, apparently by continued fault movement. The dikes both cut and are cut by red-matrix fault breccias. These dikes could not be mapped separately from their intruded fault planes. Such a dike from Ventana Canyon with similar occurrence, described as trachytic, was dated by Shakel (1974, as reported by Reynolds and others, 1986) at 21.0 $\pm$ 0.3 Ma. This age is based on whole-rock K-Ar, probably an adequate technique for these dikes.

Lamprophyre. Dikes of lamprophyre, mostly altered, occur in localities scattered through the study area excepting the northern part (pl. 1). In many dikes, sure identification as lamprophyre is made impossible by the alteration. A relatively fresh dike in the main range consists of biotite and clinopyroxene in a microlitic groundmass. The dikes generally cut mylonitic fabric in rocks of the forerange and are undeformed with chilled margins. Most dikes of the array trend NNE, and are continuous with an arc of basin-range faults to the north (pl. 1).

SURFICIAL UNITS

Deposits presumably of Quaternary age that reflect accumulation on the present land surface are widespread, but were mapped (pl. 1) only where they are extensive and thicker than local small-scale relief. All the bodies shown were examined in the field; others undoubtedly exist.

Colluvial and Landslide Deposits

Due to a possible continuum between colluvial and landslide deposits, they were mapped together (pl. 1). End-members are described along with an important questionable example.

At one extreme is an apparent landslide on south-facing slopes of Sabino basin (pl. 1), where lithologic boundaries have been translated and somewhat jumbled, apparently without much rotation of constituent parts. This landslide has a smooth land surface because it is not resistant to erosion.

True colluvium typically occupies locales of drainage convergence and (or) lesser gradient on or below steep slopes. Blocks commonly up to 10 m may show some rounding. Where mapped, these bodies are sufficiently thick that subsequent drainage has not incised through them. A large (0.5 km^2) colluvial fan at the head of the Atchley drainage (pl. 1) is at least 20 m thick, and probably owes its existence to relatively inactive downcutting in this ailing drainage segment.

Intermediate and somewhat enigmatic in character are large surficial bodies on the east wall of lower Sabino Canyon (pl. 1), which consist of rotated and unrotated angular blocks, poorly sorted in size, ranging up to about 20 m. These have accumulated on steep slopes below Thimble Peak and Saddleback ridge, and are being dissected by modern gullies. In places they form the spurs between gullies. Thus they are not colluvial fans of the most recent geomorphic configuration, and may have formed either as colluvial fans of a preceding configuration, or as landslides in earlier stages of the present one. The toes of these masses rest on a bedrock shelf that may represent a former base level. Historic landsliding is said to have remobilized a small part of one of the masses.

Alluvium

As mapped, alluvium includes both valley-bottom deposits formed during the current geomorphic cycle, and deposits stranded on terraces, incised by modern drainages. Locally, these types of alluvium are separated on plate 1. In the area between Peppersauce and Stratton washes in the northeastern part of the map area, I have included relict alluvial fans in this unit, whereas W. R. Dickinson (oral commun., 1997) concludes that they are mostly of Quiburis age. Alluvium throughout the study area consists largely of gravel, up to cobble- or boulder-size. The larger areas of alluvium occur on the margins of the study area, i.e. in the Tucson and San Pedro basins.

METAMORPHISM AND RELATED FABRICS

Rocks of the study area have undergone metamorphism of five types, 1) Precambrian metamorphism that affected only Pinal Schist, 2) Laramide dynamothermal metamorphism, 3) contact metamorphism related to Leatherwood granodiorite and Wilderness granite, 4) mid-Tertiary mylonitization, and 5) mid-Tertiary contact metamorphism resulting from detachment faulting and intrusion. No high-grade regional metamorphism has affected the area; metamorphic fabrics are more obvious than metamorphic mineral assemblages except in thin zones around plutons.

PRECAMBRIAN METAMORPHISM OF PINAL SCHIST

In the Pinal Schist, described above, the development of metamorphic fabric is rather patchy. In some areas, especially of rhyolitic facies, foliation is inconspicuous. In most areas of metasedimentary rocks, a steep phyllitic foliation is present. Metamorphic porphyroblasts are not common. Mineral assemblages are those of the lower greenschist facies

Pinal can properly be called a schist only in the belt of discontinuous Pinal outcrops (pl. 1B) that extends from the window in Alder Wash on the north (fig. 11), to Kellogg Mountain on the south (fig. 12). In this area, quartz segregated and muscovite coarsened to form finely banded quartz schist.

Pinal Schist locally contains two metamorphic fabrics, one Precambrian and one Laramide (locs. 2, 3). The older fabric in the Pinal is truncated by the unconformity at the base of the Apache Group (fig. 12). Mineral assemblages in the Pinal may also be either Precambrian or Laramide.

LARAMIDE DYNAMOTHERMAL METAMORPHISM

In the large area from the Catalina range crest to the Mogul fault, pre-Wilderness suite rocks show a foliation of varying strength. Near the range crest, penetrative deformation is apparent in all rocks—even Oracle Granite, diabase, or conglomerate may have original textures heavily modified (figs. 12, 16B), and argillaceous rocks such as those of the Pioneer Formation are phyllites, or locally schists. Strain is locally severe (loc. 3). Multiple fold episodes have locally resulted in refolded metamorphic fabrics (fig. 70, locs. 17, 95). Tectonic transport indicators are generally oriented east-west to ESE-WNW.

South of the range crest, older fabrics are present only in screens between post-Laramide intrusions, and are commonly overprinted by younger fabrics. The inclusion zone between the Wilderness and Catnip Canyon sills locally preserves Laramide fabrics (fig. 58).

North of the crestal area, the foliation is strongest adjacent to old faults, especially the Edgar thrust, the Geesaman fault, and the faults bordering American Flag Formation. Carbonate units locally are converted to marble, and show contorted folds. Bykerk-Kauffman (1990) found that much of the fabric in these situations results from ductile behavior of calcite.

Away from such faults, penetrative cleavage is restricted to the more susceptible rocks. Rice Peak porphyry, for example, commonly shows a phyllitic sheen on its groundmass, and its altered feldspar phenocrysts are recrystallized and flattened without losing euhedral intersections with cleavage (loc. 69). Argillaceous rocks in the Pioneer Formation have locally become true slate (fig. 20, loc. 11). Rock types that are somewhat more resistant to deformation, such as Leatherwood granodiorite, may show a strong foliation in certain discrete zones, and elsewhere show a spaced cleavage or none at all. Over most of this area, foliation where apparent is nearly horizontal or dips gently northeast. Here also tectonic transport was east-west to ESE-WNW. Bykerk-Kauffman (1986, 1990) found that movement sense was mostly top to the east.

The age of these fabrics is broadly Laramide. Some relations suggest considerable complexity of Laramide deformation; for example, Leatherwood granodiorite intrudes the Geesaman fault (loc. 64) and cuts some fault-related fabric, but is itself foliated in the range-crest area. Leatherwood dikes in the Marble Peak area cut some folds in Paleozoic limestones, but are boudinaged in others (fig. 71). Leucogranites and pegmatites of the Wilderness suite, even as thin dikes, generally cut these fabrics (fig. 52, locs. 3, 4, 74), but the Spencer granite, the oldest of the suite, itself contains a weak sub-parallel fabric. This range in relations suggests that Laramide fabrics were formed from the Late Cretaceous to the early Eocene.

Bykerk-Kauffman (1990) considered much of the fabric syn-Wilderness. I find that Laramide fabrics predate most Wilderness intrusions. Intrusion of Wilderness granite apparently correlates with crenula-

Figure 70 Recumbent fold in Proterozoic metadiabase on Kellogg Mountain near the range crest (loc. 95). Note the folded lineation. Outcrop about 2 m high.

tion of earlier foliations along the range crest, as described below.

Some rock bodies foliated in the Laramide show an apparent gradation with those metamorphosed by intrusion of Leatherwood granodiorite. In such cases, the combination of compressive stress and high temperature apparently produced higher-grade mineral assemblages in plastically deformed rock. An example is provided by the Paleozoic limestones both over and under the Leatherwood sill, which probably were affected both by heat from the sill and regional deformation. The result was thoroughly marbleized limestone showing moderate- to high-grade mineral assemblages throughout, and intense development of small- to outcrop-scale folding (fig. 39). Intrusion of Wilderness granite also increased local metamorphic grades, but generally produced unfoliated mineral assemblages.

In most parts of the study area, the effects of intrusion and regional foliation can be kept separate, either through distribution or relative timing. For example, fabrics that cut the Leatherwood granodiorite most postdate Leatherwood-related skarns. In those areas without nearby plutons or pluton-related plumbing systems, the mineral assemblages are low grade, most commonly sericite-chlorite-epidote, or lower greenschist facies. Near the range crest, biotite-zone or upper greenschist facies conditions are widespread. Garnet- or staurolite-bearing rocks occur, but apparently reflect Wilderness granite intrusion (Bykerk-Kauffman, 1990) or earlier Leatherwood granodiorite intrusion, respectively. Rice Peak porphyry seems not to have produced thermal aureoles, and consequently the area north of the Geesaman fault shows only low-grade assemblages. Janecke (1986) found an irregular distribution of metamorphic grades in Abrigo Formation near the Geesaman fault.

For the time of Rice Peak venting through American Flag Formation accumulation, these Laramide paleosurface indicators can be reconstructed from their current outcrop areas southward to calculate the depth of Laramide burial in foliated rocks (correcting for younger structural features). Figure 47 shows paleodepths ranging from about a kilometer north of the Geesaman fault to about 4.1 km at the range crest. Probably the paleodepths at the range crest could have been as much as about 5.5 km without a need for structures that were not observed. Henceforth this range of paleodepths at the range crest will be used in estimation of paleodepths elsewhere (in conjunction with inflation by younger sills). The paleodepths compiled for the time of Leatherwood intrusion may not be representative of the entire Laramide time interval.

Such modest paleodepths in an area of penetrative ductile fabrics, tight to isoclinal recumbent folds, and greenschist-facies metamorphic assemblages suggest high geothermal gradients. However, the high-grade contact-metamorphic assemblages associated with the extensive Leatherwood granodiorite intrusive sheet at shallow paleodepths themselves suggest high geothermal gradients.

Figure 71 Deformed Leatherwood granodiorite (Kl) in Horquilla Limestone (PPh), southeastern face of Marble Peak (loc. 96). Outcrop about 5 m high.

INTRUSION-RELATED METAMORPHISM

Intrusion of Leatherwood granodiorite produced the highest-grade metamorphic assemblages in the study area. Included in this intrusion-related metamorphism are both true contact-metamorphism, i.e. skarns and hornfelses, and adjacent metamorphism in foliated rocks at high ratios of temperature to pressure. Contact metamorphism probably preceded the development of foliation in most rocks, as the main plutonic source of heat, the Leatherwood granodiorite, is itself foliated.

Skarns and Related Rocks

Calc-silicate skarns formed in narrow zones along intrusive contacts of Leatherwood granodiorite with carbonate host rocks. In the Marble Peak district, the best-developed skarn deposits are in the Abrigo, Martin, and Escabrosa Formations. Commonly the skarns show prograde garnet-pyroxene assemblages and retrograde or metasomatic epidote-hydrous mineral-sulfide assemblages that may cut the earlier assemblage. Magnetite is locally abundant. Here I emphasize the prograde components; the retrograde components are described with their mineral deposits.

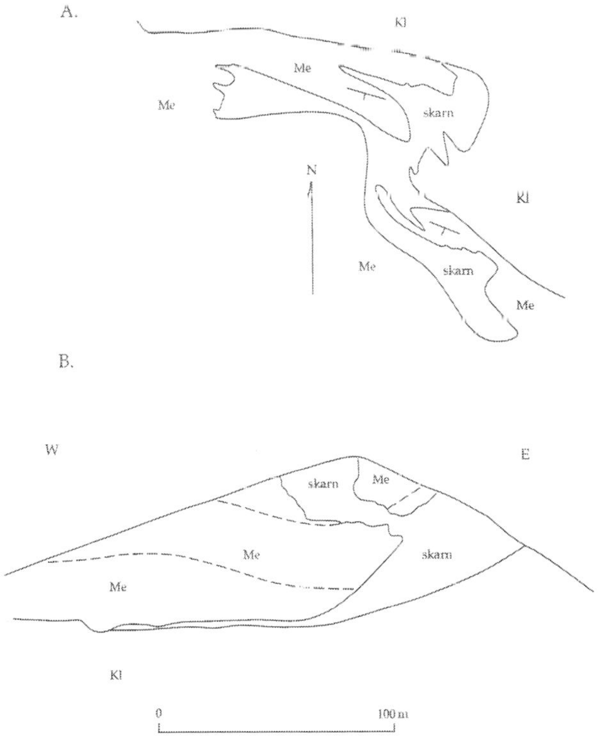

Figure 72 Christmas-tree skarn bodies on intrusive Leatherwood (Kl) - Escabrosa (Me) contact in different parts of Daily mine (loc. 149), from Peterson and Creasey (1943). A. Map view. B. Cross-section view.

Not all suitable intrusive contacts show much skarn. In detail, the larger individual skarn bodies are Christmas-tree-shaped, rooted on the contact but projecting upward 10 m or more into carbonate host rocks (fig. 72). Many such bodies are localized along pre-Leatherwood faults. Skarn mineralization along the base of the Leatherwood sill is less common than along the top even where the host rocks are the same. Peterson and Creasey (1943) observed skarn along the top but not the bottom of a subsidiary sill of Leatherwood granodiorite in the Daily mine area, described under mineral resources. Skarns along the steep-sided southeastern margin of the Leatherwood sill have been removed by erosion in Alder Creek where the margin is against carbonate rocks, and skarns were not observed toward its headwaters where the steep side is against less calcareous Precambrian rocks.

Wood (1963) found that skarns in calcareous host rocks contain diopside and minor sanidine within about 1 m of intrusive contacts, and amphiboles up to 4 m from the contact. The most common garnet he found to be andradite. Where the Leatherwood intrudes Paleozoic sandstone, Wood found diopside hornfels developed as much as 15 m from the contact. Other minerals in contact zones were epidote, amphiboles, and porphyroblastic microcline.

Some porphyritic sills related to Leatherwood granodiorite show related skarn-type metamorphism in the area east of Alder Creek. One body at the head of Atchley and Edgar drainages has produced an interval of coarse diopside about 100 m thick in lower Abrigo Formation, even though separated from the overlying sill by quartzite of the Southern Belle Member (loc. 97). Diopside forms zones parallel to bedding and upright cauliflower-shaped masses up to 1 m high. Locally this sill has recrystallized Mescal Limestone also (loc. 98). Retrograde amphibole is present in the skarn zone.

High-temperature Aureoles

Larger volumes of rock contain mineral assemblages that reflect nearby (mostly underlying) intrusion. Detailed study of these assemblages has not been conducted for this study.

The highest-grade assemblages are due to Leatherwood intrusion. At Marble Peak, Escabrosa Limestone and Martin Formation contain wollastonite (locally pseudomorphs of wollastonite) within about 100 m of intrusive Leatherwood contacts, and coarse white marble with talc partings extends more than 500 m above this contact in Martin, Escabrosa, Horquilla, and Earp Formations. Below the Leatherwood sill, the same formations have also been recrystallized to marble. In the Red Ridge area of the range crest, Leatherwood intrusion has recrystallized the overlying Abrigo Formation and the Mt. Lemmon unit (loc. 40) to staurolite-andalusite and/or garnet-amphibole assemblages. Bykerk-Kauffman (1990) has recorded sillimanite in similar circumstances north of the range crest.

Intrusions of Wilderness granite are locally near high T/P metamorphic assemblages, but in some areas such as Lemmon Rock and Marshall Gulch, where intrusion of Wilderness granite and Lemmon Rock pegmatite-aplite have followed earlier Leatherwood intrusions (pl. 1B), the resulting metamorphism is difficult to partition. I suspect that garnet-epidote assemblages in metadiabase at Lemmon Rock (loc. 36), hornblende-epidote assemblages in Abrigo Formation atop Mt. Lemmon, and perhaps garnet in Pioneer Formation in Marshall Gulch (loc. 99) primarily reflect Leatherwood-age metamorphism. All such areas are within the distribution of the Leatherwood intrusive sheet and in three dimensions within about 200 m of it; most of them show nearby Leatherwood outcrops.

In a few areas, the assemblages apparently reflect proximity to Wilderness granite rather than Leatherwood. However, I find that metamorphic porphyroblasts other than magnetite and pyrite are uncommon in pelitic units where Wilderness suite leucogranites are the only nearby intrusives (as in the screen of Pioneer Formation between the Wilderness and Spencer sills). Silicate assemblages attributable to Wilderness intrusion are commonly actinolite-epidote-chlorite (loc. 35). Porphyroblasts of Wilderness-related metamorphic grades are most common in rocks with crenulated fabrics, suggesting that Wilderness intrusion may be coeval with the crenulation. Bykerk-Kauffman (1990) interpreted garnet-amphibole assemblages in calcareous rocks as progressing northward away from the Wilderness granite contact into tremolitic rocks, but her data are ambiguous and due in part to Leatherwood intrusion.

In one area, high-temperature assemblages were found where no intrusion is exposed. This area is in Alder Creek upstream from Atchley Canyon, where marble developed from Escabrosa Limestone contains wollastonite at stream level. Known Leatherwood to the northwest is structurally higher. This locality is probably less than 30 m (100 ft) above a splay of the pre-Leatherwood Geesaman fault, exposed in a nearby window (pl. 1). I suspect a Leatherwood intrusive at shallow depth even though none is exposed at the land surface. The fault may have channeled hot fluids to the locality.

In summary, metamorphic minerals I attribute to Leatherwood intrusion (wollastonite, diopside, staurolite, andalusite, and garnet) suggest considerably higher temperatures than those I attribute to intrusion of Wilderness-suite rocks (actinolite, epidote, chlorite, pyrite, and magnetite; local garnet). Figure 73 shows my view of burial depth history, designed for contrast with fig. 4 of Anderson and others (1988). My data are derived not from geobarometry but from geologic constraints (fig. 47 and thicknesses of Wilderness suite sills). My burial depths are even shallower than those given by Palais and Peacock (1990), and show maxima after intrusion of the Wilderness suite sills at the southern end of the map area rather than at the range crest.

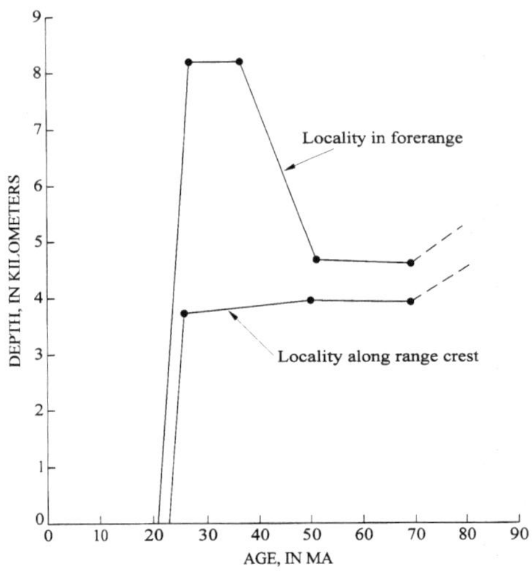

Figure 73 Diagram of burial-depth history in two parts of the study area, in the manner of Anderson and others (1988, fig. 4), but compiled from structural relations, using paleodepth figure from figure 47 and the thickness of Wilderness suite sills. This approach has the advantage of including thicknesses of the dated igneous rocks, and of allowing different depth histories for different parts of the range. Note that the greatest paleodepths were in the forerange in the mid-Tertiary. I have assumed slow erosion rates until mid-Tertiary uplift.

MID-TERTIARY MYLONITIZATION

In this section, mid-Tertiary mylonitization, a prominent feature of the forerange and south-facing parts of the main range, is addressed as a metamorphic rather than a regional-structural phenomenon. Mylonitization is dynamic metamorphism, and this equation has in part given rise to the term "metamorphic core complex" for ranges like the Catalinas (cf. Coney, 1980; Davis, 1980).

In the Catalinas, mylonitization occurs in two zones, a main zone in the forerange and adjacent parts of the main range, and a Windy Point zone higher in the main range (pl. 1B). Mylonite was produced by ductile deformation of parts of the rock (such as quartz) and brittle deformation of others (such as feldspar). The mylonites typically have a well-developed lineation interpreted as the direction of slip. Mylonitic foliation, variable in strength, commonly defines huge mullions parallel to lineation that locally show appreciable side dips. As in other mylonitic rocks, two planar fabrics are the flattening (S) surface and the shear (C) surface (fig. 74), which intersect in the study area at angles of about

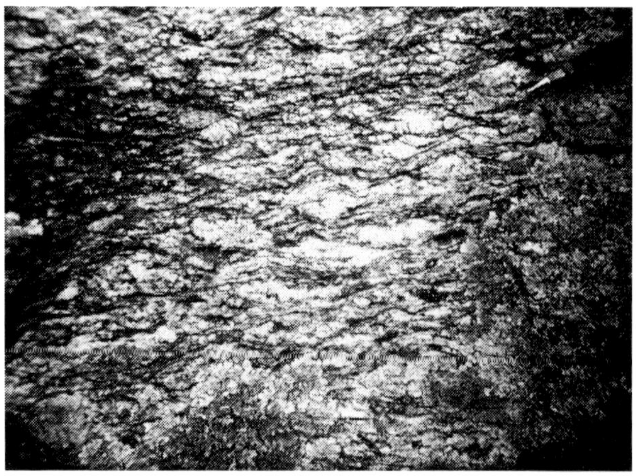

Figure 74 Top-to-northeast fabrics in mylonitized Oracle Granite, west fork Sabino Creek (near locality 100 looking SE). Shear bands (C surfaces) parallel to pencil.

15 to 40°, the lesser intersection angles corresponding to greater shear strain. Mylonitization has flattened the structure of originally angular junctions in all directions, but especially in the direction of the lineation (figs. 43 and 68), suggesting the importance of both pure and simple shear in mylonite formation. Ptygmatic folds of pegmatite dikes (fig. 75) in some domains suggest that pure shear predates simple shear. However, detailed study of mylonitic fabric was not conducted for this study.

Protoliths on which mylonitic fabric have been imposed in the study area (pl. 1B) include (from oldest to youngest) Oracle Granite, minor Proterozoic diabase, Leatherwood granodiorite, and leucocratic granites and pegmatites of the Wilderness suite. Knagge granite cuts most mylonitic fabric.

Mylonitization is concentrated in older and more biotitic rocks, particularly Oracle Granite and Leatherwood granodiorite. Sills of Wilderness suite leucogranite are somewhat less mylonitic, probably due to greater competence. This contrast is apparently responsible for a marked parallelism of mylonitic fabric to sill boundaries, a feature noted elsewhere in the region by Rehrig (1982). Parallelism was preserved when both sills and fabrics were folded. The geometry of the sills is thus a primary control of the geometry of mylonitic foliation. This combination has in turn controlled the shape of the land surface (fig. 10).

In the uppermost exposed mylonites of the Catalina forerange, the cataclastic component of mylonitization increases upward over about 100 m, as an overprint on more ductile fabrics. Both limbs of the forerange anticline show this cataclastic component in their upper parts, suggesting later-stage movement at shallower crustal levels there.

Mid-Tertiary mylonitization of a least two ages has occurred in discrete levels that converge to the southwest; this is discussed under structure. Except where overprinting and divergence occurs, the envelope of mylonitization in a given area is probably about 600 m (2000 ft) thick (cross sections, pl. 1). An approximate coincidence of topography with the mylonite zone in the Catalina forerange has made the mylonite areally extensive, however.

The base of the Windy Point mylonite zone shows two interesting geometric features. First, the base shows huge mullions of mylonitic rock (pl. 1B) about 1.5 km wide with vertical relief of as much as 250 m (800 ft). Second, S surfaces arc away from the base of the zone (fig. 76), as predicted from the larger angle between C and S surfaces in areas of less strain, but below the main mylonitic zone, the fabric becomes widely spaced rather than penetrative (fig. 76). Little-sheared zones averaging about 45 m (150 ft) thick are separated by mylonitic zones 10 m or less thick that are poorly exposed but cut primary banding. This spacing is rather regular, and corresponds to the distribution in Wilderness granite of a characteristic palisaded topography of this part of the main range (fig. 4), and a characteristic banded pattern of vegetation in air photos. These features can in turn be used to predict spacing of mylonitic bands in appropriate areas. The two features described here combine to make work along the base of the mylonite very tricky; a given non-mylonitic outcrop could either be outside a mullion, or over a spaced shear zone.

Mineral assemblages attributable to mylonitization in the study area are those of the greenschist facies, including the biotite zone. Anderson and others (1988) suggest from muscovite geobarometry that mylonitization was at a depth of about 9 km. My

Figure 75 Ptygmatic folding of Wilderness suite leucocratic intrusives in Oracle Granite, both mylonitic, along the crest of the forerange anticline, looking oblique to lineation; cliff on east side of Molino Canyon (loc. 89).

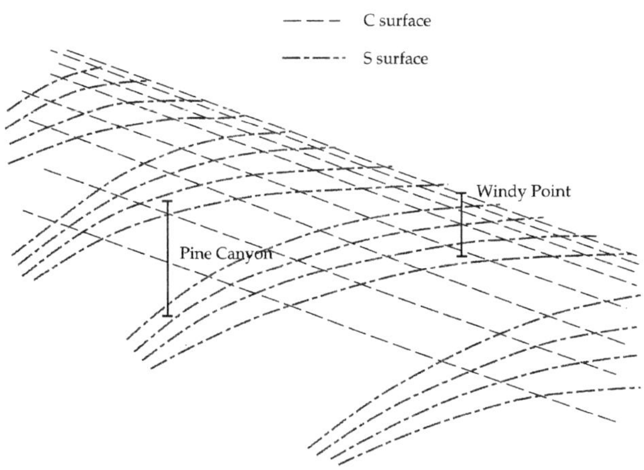

Figure 76 Diagram of arcing of foliation bundles on the bottom of Windy Point mylonite zone in the Wilderness granite sill, looking southeast. The portrayed spacing of fabric elements is proportional to fabric intensity. The vertical dimension about 400 m.

structural reconstructions suggest that mylonitization was as deep as about 8.7 to 10.1 km and as shallow as about 4.5 to 5.9 km.

The age of mylonitization is suggested by mylonitic deformation of Wilderness suite leucogranites as young as 46–49 Ma, and by deformation of Catalina Granite, about 26 Ma, by the younger (Windy Point) but not the older belt of mylonite. The average of the K-Ar dates on micas from mylonitized Wilderness suite rocks of the area, listed by Reynolds and others (1986), is also about 26 Ma. These dates are probably accurate enough to indicate that mylonitization was mid-Tertiary.

The pure shear component of early mylonitization of the study area is somewhat unexpected, but compares with mylonites of the well-studied core complexes of California (fig. 1; Davis and others, 1982). Some of these California mylonites predate mid-Tertiary extension (John and Mukasa, 1990), however, unlike the study area where mylonitization is apparently related temporally to extension. Earlier penetrative ductile fabrics were probably present in pre-Eocene rocks throughout the southern part of the study area, but I would not call these mylonitic.

MID-TERTIARY HORNFELS

In the San Manuel-Mammoth area, epidote-chlorite assemblages have formed in the upper plate of the nearly horizontal Cloudburst detachment fault (loc. 56). The distribution of these assemblages suggests they are due not to igneous intrusion but to tectonic juxtaposition of hot lower plate rock against colder upper plate rock (Force and Cox, 1992), with attendant fluid flow. Chlorite and epidote occupy conglomerate matrix, vesicle fillings, and veins. They extend above the Cloudburst fault plane about 50 to 90 m (150 to 300 ft).

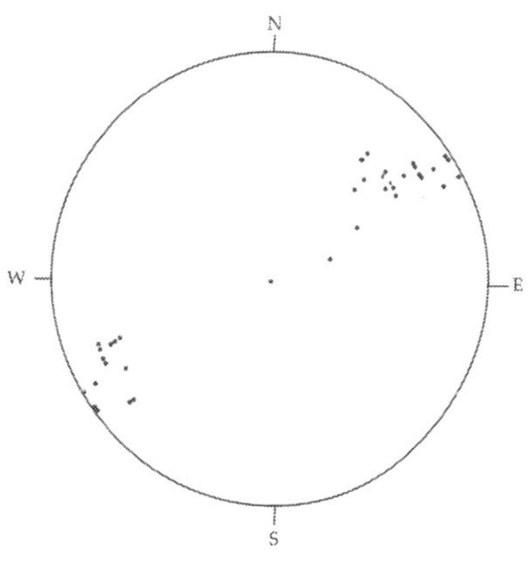

Figure 77 Stereographic projection of folded mylonite lineation in the Catalina forerange.

Replacement by these minerals is especially pervasive adjacent to the steep Mammoth vein set, where metamorphism was hastened by greater fluid flow derived from the cooling of small rhyolite intrusions.

STRUCTURES AND STRUCTURAL EVOLUTION

In this section, parts of the study area are examined individually from south to north, or from the deeper structural levels to the shallower. For each part, structures are described in order of decreasing age (to the extent this is possible), then the area is analyzed in a "backstripping" mode in which the effects of structures, rock units, and other features are progressively removed from youngest to oldest in order to visualize intermediate states.

FORERANGE AND SOUTHERN MAIN RANGE

This is the principal area where original Precambrian basement, represented by Oracle Granite, is exposed in the Catalinas south of the Mogul fault. It is also an area of voluminous leucogranite sills of the Eocene Wilderness suite, and pervasive mid-Tertiary mylonitization, both of which have obscured original relations in older rocks.

Laramide deformation

The few outcrops in this area where pre-mylonitic features are preserved are in the angle between the main and Windy Point mylonite zones, i.e. at intermediate elevations in the south-facing main range. These outcrops show strong penetrative fabrics, locally refolded (fig. 58, loc. 83) in pre-Eocene rocks. They are quite unlike mylonitic fabrics and among the strongest observed in the study area. The southern part of the study area was apparently deformed in the Laramide, but most evidence obliterated.

Mylonitization

Mid-Tertiary mylonitization is the oldest structural feature that can be recognized in most outcrops in this area. The dominant sense of shear movement in the mylonite is top to the southwest, as has been documented by many workers (Davis, 1980; Naruk, 1987). In detail, at least two ages of mid-Tertiary mylonitization each with different loci are present, as proposed by Reynolds and Lister (1990).

First stage. The oldest mylonitization, well represented in the forerange and the adjacent base of the main range, is concordant to leucogranite sills of the Wilderness suite; most deformation is focused in screens of Oracle Granite and Leatherwood granodiorite. Where not overprinted by later mylonites, the preserved thickness of this mylonite is about 600 m (2000 ft), that is, approximately from the Catalina detachment fault to the base of the zone at the Seven Falls sill (pl. 1), which shows ductile strain even in feldspar (locs. 75, 76). The cataclastic component of mylonitization increases rapidly upward in the highest exposed mylonites.

This main belt of mylonite is folded in the Forerange anticline (pl. 1) of Peterson (1968). Reynolds and Lister (1990) showed that every fabric element in the rock is folded; that is, the mylonite did not form in this folded configuration but was itself deformed. The axial plane of this fold strikes about N. 60° W. toward the western margin of the study area, and swings almost east-west toward the east (pl. 1, locs. 77, 89). The axial plane dips south or southwest due to the asymmetry of the fold, and this dip is reflected in the trace of the axial plane on plate 1.

Mylonitic foliations are irregular. For about 90 percent of the width of the fold, the limb attitudes differ less than 35°; however, on the northern margin the northeastward dip is locally quite steep (fig. 74) , even overturned (loc. 100).

Mylonitic lineation is folded such that it trends about N. 65° E. and plunges gently southwest on the southwestern limb, but trends about N. 45° E. and plunges variably northeast on the northeastern limb. These data (fig. 77) suggest a fold plunging about 0 to 20° to the southeast, but a reliable axis cannot be calculated from these data because linear features oblique to the axis of later folding may or may not define a great circle in stereographic projections, depending on the fold style.

Along its northern margin, this belt of mylonite dips under the main range. However, considerable complication where this occurs, in Sabino and Molino basins, is caused by mylonite reactivation and faulting.

Reactivation. All along the northeastern limb of the Forerange anticline, top-to-southwest fabrics of first-stage mylonite are overprinted by top-to-northeast mylonitic fabrics (as in fig. 74). Overprinting can be observed in some individual outcrops but more commonly in laterally adjacent domains (locs. 100, 102). Locally the overprinting involves mylonitization with a higher proportion of cataclasis than the first stage (north of loc. 60). The area of overprinting extends from where mylonite dips under the southern base of the main range (pl. 1B), southward to the present crest of the forerange

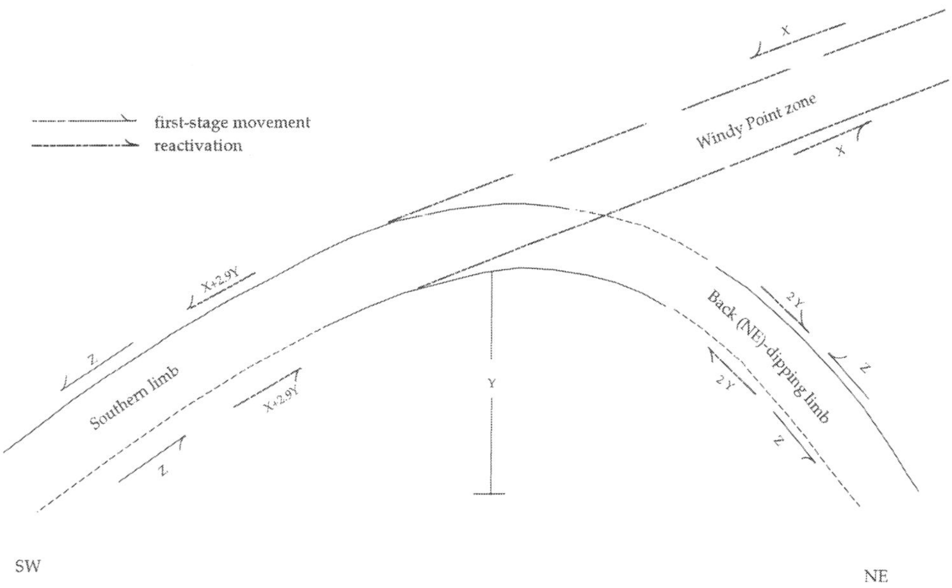

Figure 78 Sketch of relations of the two stages of mylonitization, looking northwest. Vector components x, movement on Windy Point zone; y, vertical component of mylonite reactivation; and z, first stage-movement are calculated in text under structural restoration.

anticline. I propose that uplift of the forerange during mylonitization caused folding of first-stage mylonite, and locally reversed the direction of shear such that movement became coupled as with a diapir, i.e. top-away-from-crest (fig. 78).

Ptygmatic folds (fig. 75, locs. 89, 101) of leucogranite and pegmatite dikes occur throughout the forerange, especially north from the crest of the forerange anticline. The shapes of these folds suggest flattening rather than simple shear. Those along the crest could be the result of a null zone during reactivation (if some of the leucogranites postdate first-stage mylonitization), and those to the north may result from changed motion during reactivation. However, a satisfactory history of these folds will require more work.

The reactivated mylonites have partly-ductile fabrics, but local northeast-dipping cataclasis, mostly as pervasive fracturing without brecciation or gouging, apparently records the last stages of the process in the brittle zone (fig. 79). This cataclastic component increases structurally upward toward the younger Romero Pass fault zone.

On the southern limb of the Forerange anticline, mylonitic foliation defines a subtle northeast-trending antiformal structure (the Sabino Canyon fold of Peterson, 1968) parallel to lineation. This antiform probably reflects the internal structure of a mullion, rather than true folding. That is, it is a primary feature that like lineation forms parallel to tectonic transport. The foliation pattern is mimicked by the shape of the younger Catalina detachment fault, the upper bounding surface of this mylonite. The gradual termination of this foliation pattern at the crest of the Forerange anticline suggests that the mullion is a first-stage structure subsequently destroyed on the northern limb.

The cause of formation of the Forerange anticline is not clear. It began growth in the semi-ductile zone and seems to have eventually caused *upward* migration of the locus of shear to the Windy Point zone, unlike buckling above a deep younger detachment (cf. Frost and Heidrick, 1996). The sense of last-stage movement was top-away-from-crest, as with a diapir. If the mechanism of fold growth was rebound from tectonic denudation of a crustal root, such denudation would have to predate the Windy Point mylonite zone, which is not folded within this part of the area.

Naruk (1987) proposed a shear-zone model for many of the features I have included with reactivation. This shear zone would be steep with top-to-NE sense, and cut the older mylonite. Like Reynolds and Lister (1990), I have rejected this model because it does not predict the distribution of top-to-NE fabrics correctly nor explain the Forerange anticline. Specifically, I find

Figure 79 Cataclastic mylonite derived from Leatherwood granodiorite, reactivated zone, Molino Basin (north of locality 60).

that top-to-SW fabrics dipping NE are common near stop 3 of Naruk and Bykerk-Kauffman (1990), i.e. their shear zone had an earlier history of mylonitization.

Windy Point zone. A spatially distinct locus of mylonitization is high in the main range, where mylonitized Wilderness granite has been much-visited at Windy Point (loc. 103). Shear sense is everywhere top-to-southwest; C-S angles are about 25° in penetratively deformed rock. Such mylonite forms a zone of discrete outcrop areas across the south flank of the range crest. The zone is best visualized as having been continuous and dipping very gently southwest, but with huge northeast-trending mullions along its base. This basal surface crops out where structurally low mullions intersect topographically high land surfaces (pl. 1B). Thus the Windy Point zone of mylonite barely grazes the main range at the present level of erosion.

The mullion that includes Windy Point itself is about 1.5 km wide and shows relief of about 250 m

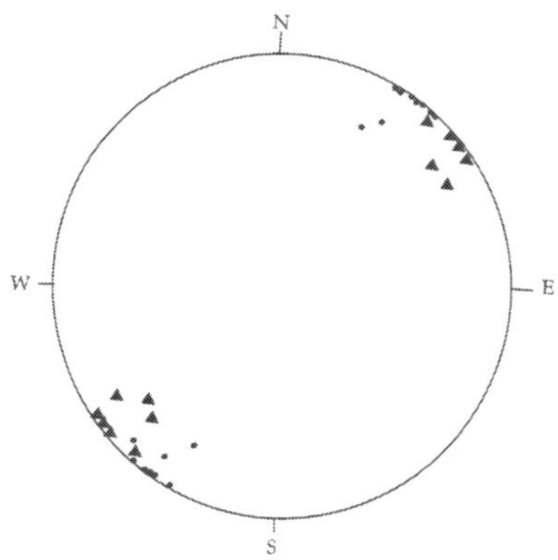

Figure 80 Stereographic projection of mylonite lineation in the Windy Point zone. Note difference in trend of the eastern (dots) and western (triangles) parts of the zone.

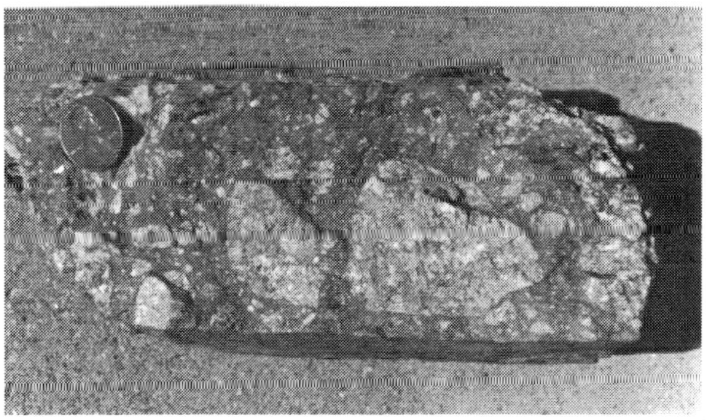

Figure 81 Specimen of typical red-matrix breccia of the Sabino Canyon fault.

(800 ft; pl. 1B). In a few areas like Rose Peak, mullion margins can be seen in cross section. Mylonitization becomes more feeble toward the northeast, so that at Leopold Point (loc. 104), where the mylonite is last seen, the angle between C and S fabric elements is over 30°.

The Windy Point zone deforms Catalina Granite near Shovel Springs, west of the study area. Elsewhere, Catalina Granite cuts first-stage mylonite, showing the relative age of these mid-Tertiary features. Knagge granite mostly cuts the Windy Point mylonite zone (pl. 1B), but a mylonite zone about 10 m thick is exposed within Knagge granite at Barnum Rock (loc. 105) and Leopold Point (loc. 104).

Mylonite lineations in the Windy Point zone parallel the mullions on its base. Plate 1 and figure 80 suggest that lineation and apparently mullion elongation converge to the northeast.

The development of spaced mylonitic zones in the footwall of the Windy Point zone has been described above (fig. 76). A cross section through the spaced mylonitic fabric below the Windy Point zone is best exposed at locality 106. This locality also displays an intersection of primary banding and mylonitic fabric.

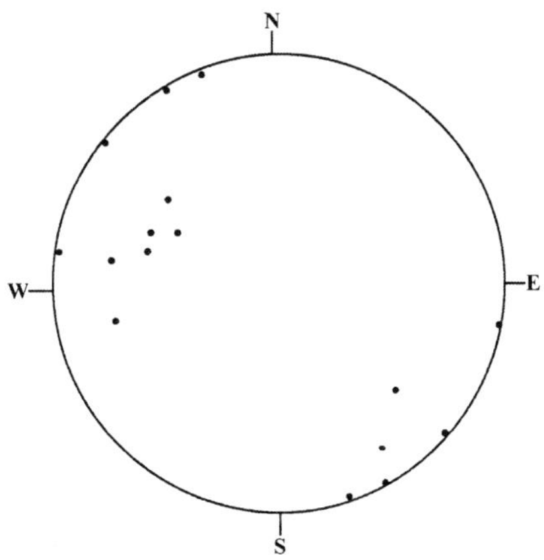

Figure 82 Stereographic projection of poles to planes of Sabino and Bear Canyon faults.

Intersections of the Windy Point zone with mylonites of the forerange are not exposed, due to erosional excavation of the Sabino and Molino basins (pl. 1B). Structurally intermediate rock is most commonly protoclastic, with variably developed lineation that plunges gently northeast, as might be expected from less-strained regions of either the Windy Point zone or the back-dipping limb of the main mylonite zone (fig. 78). Low angle truncations in proto-mylonite fabrics are locally apparent below the Windy Point zone, and may record superposed shear episodes.

Projection of the Windy Point zone suggests that it could have merged tangentially at the crest of the Forerange anticline, especially where dip-slip movement on the Romero Pass fault zone was modest. Probably, the Windy Point zone represents a late locus of top-to-southwest shear, during growth of the Forerange anticline, on a plane that remained efficient for such shear (fig. 78). Motion on the Windy Point zone is thus likely to be coeval with that of reactivation of mylonite on the northeastern limb of the Forerange anticline. The Catalina detachment fault, now eroded over the Catalinas, probably followed the Windy Point zone when exhumation brought the zone into a brittle shear regime.

Sabino and Bear Canyon faults

Parallel to the arcuate courses of both lower Sabino and lower Bear canyons are long curved faults that typically crop out on the southeastern canyon wall. The faults are steep, and are normally expressed as red-matrix breccias (fig. 81) about a meter wide, or as red gouge. Locally they are intruded by biotitic granitoid dikes (loc. 107). Slickensides where exposed suggest down-dip movement.

The attitude of both faults swings from a strike of about N. 60° E. near the canyon mouths to due north at the intersections with the Romero Canyon fault zone. The Sabino Canyon fault, for which there is more information, also shows a change in dip, from about vertical at the south end to about 60° eastward at the north.

The curved traces might be thought to result from folding on the Forerange anticline. However, in three dimensions the attitudes of the fault planes (fig. 82) are not closely related to the axial plane of the Forerange anticline (pl. 1). Four other lines of reasoning suggest that the Sabino and Bear Canyon faults are not folded: 1) The faults are entirely brittle features, and we have seen that formation of the Forerange anticline was initiated in the semi-ductile zone; 2) The degree of limb rotation for mylonite (30°) through 90 percent of the width of the Forerange anticline is less than that for the faults (50°) over the same width, though the older mylonite should show at least as much deformation as the faults, and the attitude of the faults is inefficient for showing the deformation; 3) The locations of greatest flexure for the faults and the fold are not the same, and 4) The movement geometry of the faults, discussed below, suggests they followed folding. I conclude that the curved faults represent a trajectory of minimum resistance through a previously folded pile of extremely inhomogeneous rocks (i.e. leucogranite sills and screens). To some extent the fault traces are parallel to lineation trends across this fold.

The pattern of fault offset is unusual. Both faults offset the Seven Falls, Gibbon Mountain, and Thimble Peak sills (pl. 1) in such a way that the block between the faults, i.e. between Sabino and Bear Canyons, has scissored (fig. 83) with the northern part down (fig. 55; loc. 79) and the southern part up (loc. 76, 77). The axis of rotation is northeast of the axis of the Forerange anticline, at about the position of Thimble Peak, but axis location is complicated by a cross-fault at about this position. Due to multiple fault strands and indeterminate offsets, I cannot give precise separation magnitudes, but the northern part of the Sabino-Bear block may be down about 30 m (100 ft) and the southern part up about 150 m (500 ft), with the blocks on either side little-rotated with respect to each other (fig. 83). Thus the Sabino-Bear block can be likened to a rotated stone in an arch.

The age of this faulting is not known from evidence within the study area; however, the brecciated trachyte dike in a similar structure, dated by Shakel (1974) as about 21 Ma, gives the age of such faulting in Ventana Canyon to the west.

Romero Pass Fault Zone

A fault system that crosses the entire width of the study area occupies the string of basins, saddles, and valleys that separates the Catalina forerange from the main range (figs. 5, 54). The name Romero Pass fault zone was mentioned by McCullough (1963), named from the high point in its path to the northwest. The zone continues to the west end of the range. Its extent to the east is uncertain.

Like the Sabino and Bear Canyon faults, the Romero Pass fault zone characteristically contains red-matrix breccia or red gouge. At several places along its length the fault plane is intruded by biotitic granitoid dikes. Slickenside plunge ranges from vertical to 33° east.

Study of this fault zone is complicated by the fact that it follows the locus of the cataclastic component of mylonite reactivation on the northeastern limb of the Forerange anticline. However, in my experience shattered mylonite dips to the northeast in seven of nine observations, whereas gouge and breccia of the Romero Pass fault zone is vertical or dips steeply southwest in ten of thirteen observations. A few individual outcrops show both features (loc. 108), with south-dipping gouge or breccia cutting north-dipping cataclastic mylonite.

Offset on the Romero Pass fault zone is difficult to measure because no rock units can be traced with certainty across its main strand in the study area. However, in the west fork of Sabino Creek, the offset base of the Wilderness granite sill can be traced across the main strand (fig. 60). Its offset from about 4500 ft (1370m) elevation near Hutches Pool on the northeast side, to about 5300 ft (1610 m) on the southwest side, implies separation of about 240 m (900 ft), up on the southwest. However, this measurement is made over a horizontal distance of 3 km. A northern strand of the Romero Pass fault zone in the study area offsets the base of the Catnip Canyon banded granite at least 200 m (650 ft), south side up.

Relations on the main strand in the study area, based on sill thicknesses measured elsewhere in the study area, suggest offset of about 1200 m (3900 ft), up on the south, in the Molino basin area. Indeed geologic relations along the fault trace everywhere suggest greater offset toward the east; figure 84 shows the apparent stratigraphic separation across the Romero Pass zone. However, in some places the observed offset would actually include some component of mylonitic reactivation, which was also up on the south. The sense of motion, combined with the predominant south dip of the fault, implies that the Romero Pass zone is a reverse fault. It may also have had some right-lateral movement, based on slickenside orientation. Perhaps it formed by continued uplift of the forerange in the brittle zone

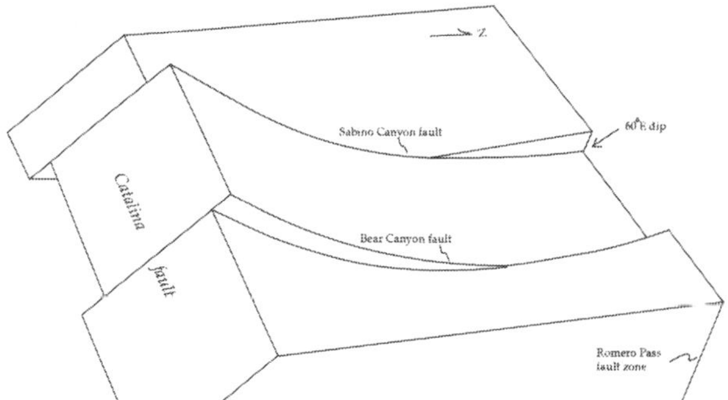

Figure 83 Diagram of block rotation due to Sabino and Bear Canyon faults in the forerange.

after mylonitization and north-dipping cataclasis ceased.

The relative ages of the Romero Pass zone and the Sabino and Bear Canyon faults is ambiguous in detail, but in broad map pattern the Romero Pass zone truncates the other faults. At the intersection with the Sabino Canyon fault (loc. 109), the Romero Pass gouge seems to cut the Sabino Canyon gouge, but relations are confused by an adjacent landslide. The relation of the Bear Canyon and Romero Pass fault zones is confused by multiple strands of both faults. However, no northern continuation of the Sabino or Bear Canyon faults has been found past the Romero Pass zone. An apparent northern continuation of the Bear Canyon fault suggested by air photographs proved to be only a dense joint set.

The similarity of fault rocks and intrusions in them suggest that all three faults may be contemporaneous. Scissors movement of the Sabino-Bear block in the manner of figure 83 requires a free surface in the location of the Romero Pass fault zone. Opposite the block, i.e. in the east fork of Sabino Creek, total offset on the Romero Pass fault zone should be about 30 m (100 ft) less than for adjacent blocks on the south side.

Catalina Detachment Fault

The southern boundary of bedrock that forms the Catalina range is the Catalina fault, which dips gently to the south. Some authors have treated this fault as the most important structure of the study area. Immediately above the Catalina fault in the study area are deformed gravels of the Pantano Formation, commonly dipping into the fault. Below it are mylonites developed from the Oracle Granite and Wilderness suite leucogranites, especially the Gibbon Mountain sill. Structurally upward within these mylonites, within about 30 m of the projected fault, cataclasis increases, so that the rock becomes shattered and saccharoidal, and locally folded. Mylonitic lineation is commonly obscured.

The fault itself is not well exposed in the study area, because it is covered by younger terrace gravels. Pashley (1966) and Davis (1980) have described the fault along strike nearby. It consists of indurated fine-grained mylonite and breccia, which pass gradationally downward into the cataclastic mylonite described above, but sharply cuts upper-plate rocks. The dip of the fault, mostly projected from along strike, is about 20° to the south. Thus the fault is approximately parallel to

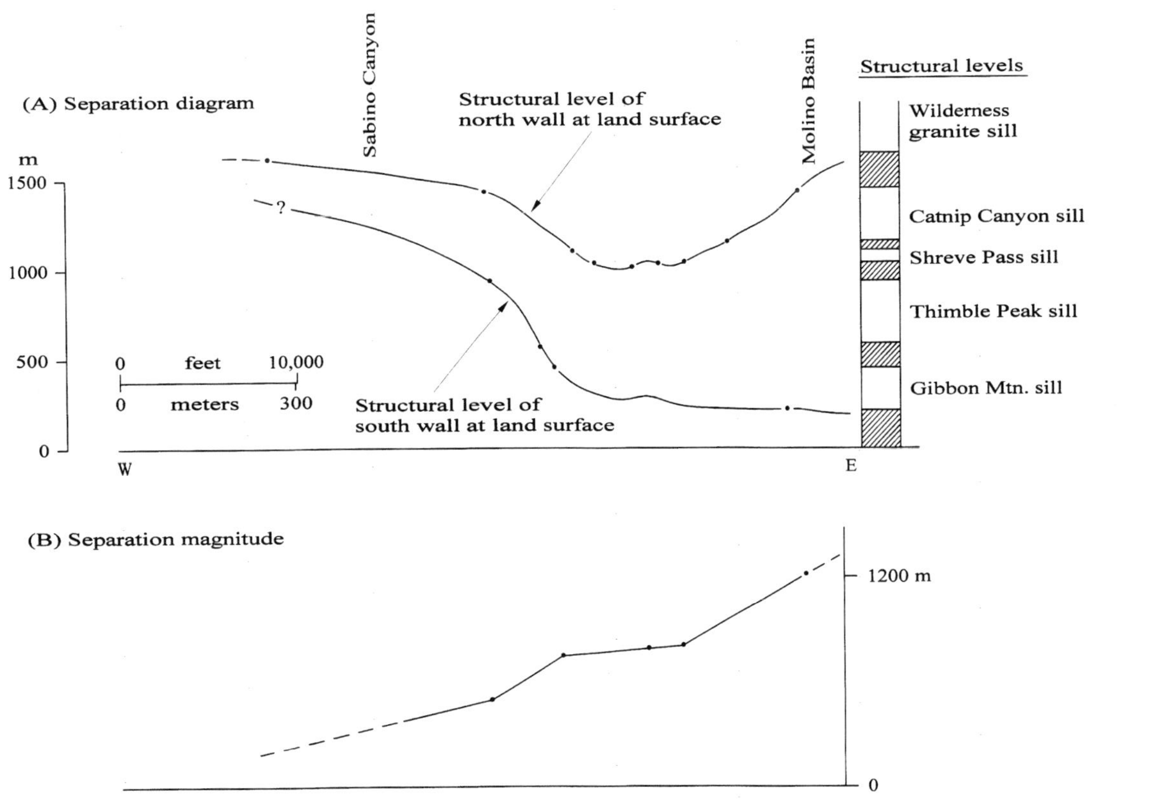

Figure 84 Structural separation diagrams for Romero Pass fault zone. The base of the Wilderness granite sill may drop to the west, and(or) dip-slip motion may scissor near Romero Pass (west of diagram). A. Structural level of each wall at ground surface; B. Separation magnitude.

mylonitic foliation in the lower plate. Commonly, springs or mesquite groves are alined along its trace in the study area.

The trace as shown in plate 1 implies that the Catalina fault postdates the Sabino and Bear Canyon faults. That is, the trace of the Catalina fault is sinuous, due to its low dip and to mullions on its surface (Pashley, 1966), but seems to cut the traces of the canyon-parallel faults (pl. 1, here after Banks, 1974, and Pashley). Offset on the Bear Canyon fault is unlikely given the straight trace of the Catalina fault, which though poorly exposed is reasonably well constrained. A swing in strike of the trace near the intersection with the Sabino Canyon fault is possible, but is better explained by concordance of the fault with a well-known mullion in mylonitic foliation (Pashley, 1966; Peterson, 1968; Banks, 1980).

Movement on the Sabino and Bear Canyon faults has thrown the block between them (fig. 83) up at least 150 m (500 ft) near the intersection with the Catalina fault (because throw increases to the southwest). Using a dip of 20° for the Catalina fault, this block would be offset at least 420 m (1370 ft) into the Tucson basin if the Sabino and Bear Canyon faults postdated the Catalina fault. This apparently has not occurred.

The younger age of the Catalina fault is consistent with other age relations. If movement on the Sabino and Bear Canyon faults occurred about 21 Ma, they probably predate uplift along the Catalina fault based on 1) fission-track dates on apatite (average about 19 Ma toward the southern end of the study area; summarized by Reynolds and others, 1986), 2) K-Ar dates on illite in Catalina fault gouge (20–21 Ma; Shafiqullah and others, 1988), and 3) mylonitic leucogranite clasts in Tinaja beds (age limits 24 to 12 Ma).

In detail then, the Catalina detachment fault must depart from mylonite of the forerange, as the Sabino and Bear Canyon structures offset first-stage and reactivated mylonite (enough apparently to expose the base of the mylonitic zone) but not the detachment fault. The detachment beveled projecting mylonite. The Catalina fault is not exposed in the range itself, but in projecting it we can expect minor and perhaps major departures of the detachment from mylonite. A case has been made above for detachment above a projected continuous zone of last-stage mylonite of which the Windy Point zone is a segment (fig. 78). This implies large-scale departure of the Catalina fault from first-stage mylonite, which dips under the main range.

The suggested temporal and local spatial departure of mylonite and detachment fault is noteworthy, because inferred spatial correspondence of these features here and elsewhere has led to a view of detachment as occurring as shear is brought from a ductile zone into a brittle zone (Davis and others, 1986; Spencer and Reynolds, 1989). Departures of mylonite and detachment are common in the California segment of the belt of metamorphic core complexes, however (Davis and others, 1982; John and Mukasa, 1990). I would suggest that mylonite belts may form at any time and place in extending middle crust, and that some segments may preferentially become loci of detachment.

Structural Restoration

Restoration of the features of this area, progressing from youngest to oldest, results as follows:

Catalina fault. Restoration of the Catalina detachment fault (17 to 19 Ma) places a structural section similar to the exposed section atop the fault if detachment roughly followed last-stage mylonite. Since this fault is exposed only at the foot of the range, this restoration is "up in the air" and does not change exposed relations. The missing rocks include an unknown amount of last-stage mylonite that presumably lay above the detachment before faulting. The upper-plate rocks are probably now in the Tucson basin; they do not appear to be those forming the Tucson Mountains, where a 4.5 km pre-mylonite section (Lipman, 1993) has too few similarities to rocks of the Catalinas.

Romero Pass fault zone. It might seem that restoration of apparent reverse movement on the Romero Pass fault zone should begin by tilting the fault plane to a normal configuration. Note, however, that we have no justification for doing so at this stage; tilting on the forerange fold predates the fault.

Restoration of slip on the Romero Pass fault zone (about 21 Ma) depresses the forerange portion of the study area from 200 to 1200 m (fig. 84). It was at this stage that the bottom of the pile probably had a burial depth of about 8.8 to 10.2 km (4.1 to 5.5 km of supracrustal load + 3.9 km of Wilderness sills + 0.7 km of Oracle-Leatherwood screens). The configuration of mylonite zones at this time was that shown in figure 78. The shallowest mylonitization (on the Windy Point zone at Leopold Point) was at a reconstructed depth of about 4.5 to 5.9 km.

Mylonite reactivation and Forerange fold. Restoration of the latest stage of mylonitization and attendant folding (21 to 26 Ma) involves simultaneous restoration of three components: 1) the Windy Point mylonite zone, 2) the reactivation of the main mylonite zone in the north-dipping limb of the Forerange fold, and 3) the Forerange fold itself. Figure 78 shows the relation of these three components. The Forerange fold apparently was initiated by coupled slip along its limbs in the semi-ductile zone. In addition, the Windy Point zone and the southern limb of the fold were the locus of top-to-southwest shear.

The minimum local amplitude of the Forerange fold is about 1100 m, i.e. from about the top of the Thimble Peak sill into the Seven Falls sill. The amount of slip on reactivated mylonite on its north-dipping limb is difficult to estimate, because strain is not penetrative, but where the dip is 30° or greater, I estimate that mylonite showing top-to-northeast fabrics, with average C-S angle of 20°, aggregates 250 m in thickness. The implied slip is 700 m^2, for which the vertical component is about 350 m. These figures suggest that perhaps a third of the fold amplitude is the result of growth during mylonite reactivation. The remainder presumably occurred in the brittle zone, as suggested by the cataclastic overprint on these rocks. Thus in figure 78, the vertical component of reactivation (y) is about 350 m. The persistence of some top-to-northeast shear up to the approximate crest of the fold suggests that this crest was a null point, and that the present apparent crest was the original crest.

The magnitude of top-to-southwest shear of this age generation is best estimated on the Windy Point zone. The exposed thickness of this zone is only about 100 m; an unknown amount has been eroded. The smaller C-S angles average about 25°. This suggests a slip magnitude of about 200 m for exposed mylonite. Thus in figure 78, x= 200 m or more.

Top-to-southwest shear of this generation in the main zone where it dips about 20° in the south limb of the fold is the sum of shear on the Windy Point zone (x), and shear required to produce the ductile component of structural relief (y/sin 20°), or x + 2.9y on figure 78. Values calculated above suggest this shear magnitude is about 930 m or more. This value is separate from that for later cataclasis, including that on the Catalina fault.

Restoration of the fold and its attendant features produces a 8.8 to 10.2 km thick pile of sub-horizontal rocks, except where Laramide deformation was severe. First-stage mylonite becomes a sub-horizontal feature also, at about 8.1 to 10.1 km depth. Topographic relief of the range at this stage is unclear.

First-stage mylonite. Restoration of first-stage mylonitization (about 26 Ma) involves a translation of most of the rocks of the study area back to the northeast relative to the Seven Falls sill, and does not change any of the relations between bodies of exposed rock. The indicated magnitude of movement on mylonite of the present southern limb is about 1650 m, based on mylonite thickness of 600 m and a C-S angle of 20°. However, from this figure we must subtract strain due to later movement, calculated above as 930 m or more, giving 720 m or less for first-stage movement (z in fig. 78). This modest figure suggests that restoration of first-stage mylonite implies no great paleogeographic change.

Wilderness suite intrusion. Restoration of the situation before intrusion of the Wilderness suite leucogranite sills (46 to 55 Ma) removes 3.9 km of structural thickness from the rock sequence, all but 400 m of it below the Apache Group unconformity. At this time, the lowest rocks we can see at present were buried about 4.8 to 6.2 km. The Leatherwood granodiorite formed several stocks strung along a sill at about 4.6 to 6.0 km depth. The entire package of exposed basement rocks below the Apache Group is only about 700 m thick.

Laramide situation. The few outcrops that preserve Laramide fabrics suggest penetrative deformation in this part of the study area. Restoration of Laramide features is tenuous, because these features are obscured except for Leatherwood stocks of patchy distribution and the Leatherwood sill, probably no more than 100 m thick. Laramide paleodepths for this area, about 5–6 km, are the greatest in the study area (fig. 73).

Restoration for pre-Laramide conditions is of interest partly because of the distribution of Precambrian metadiabase. The structural levels presently containing metadiabase, upward from the base of the Gibbon Mountain sill, would have been only 500 m or less below the base of the Apache Group. Howard and others (1991) noted that in nearby intact crustal sections, diabase sheets intrusive into basement are most common within a kilometer below the Apache Group.

RANGE CREST TO GEESAMAN FAULT

This is an area occupied mostly by northeast-dipping Proterozoic and Paleozoic supracrustal rocks that originally lay on Proterozoic Oracle Granite, but now sit structurally on Eocene Wilderness suite leucogranite. The area has been extensively intruded by Cretaceous Leatherwood granodiorite. Remarkable older structural features are present: 1) tight recumbent folds along the range crest that predate Wilderness suite granites, 2) thrust faults, some of pre-Leatherwood age, and 3) steeper faults intruded in some cases by Leatherwood granodiorite and in others by Wilderness suite leucogranite. These old structures require the type of treatment shown in figure 47 for reconstruction of the pre-Laramide situation.

Folding

Range-crest area. Physiographically, the position of the range crest corresponds to several synclines that range in character from a broad open syncline in Paleozoic rocks at Mt. Lemmon to a tight syncline in Proterozoic rocks at Mt. Bigelow. These synclines may have little relation to each other, but are a factor in the gentle topography along the crest.

Along the crestal area of the range in the Mt. Bigelow area, the Apache Group has been thrown into tight recumbent folds with a gently northeast-dipping axial-plane foliation. The structures were studied and mapped in detail by Waag (1968). Several generations of folds are present, and refolded fabrics are common (fig. 70, locs. 17, 95). Elsewhere in the study area, refolded folds are unique to the Pinal Schist.

Waag regarded all the folding stages as parts of a single syn- to post-metamorphic episode. Crenulation axes and intersection lineations trend mostly WNW parallel to axes of small folds (pl. 1). Elongated strain markers such as conglomerate clasts (loc. 3) trend WNW also, but some mineral lineations plunge gently northeast. Thus fold axes and tectonic transport directions are commonly subparallel. Where the elements diverge, as along Sunset Trail (loc. 5), the fold axes are the less regular.

Superposed folding may be responsible for some observed parallelism of tectonic elements, but sheath folding (of Cobbald and Quinquis, 1980), i.e. rotation of fold axes into tectonic transport directions, is consistent with all the observations and is probably the general explanation.

Bykerk-Kauffman (1990) has worked with the same deformation system north and east of Waag's area.

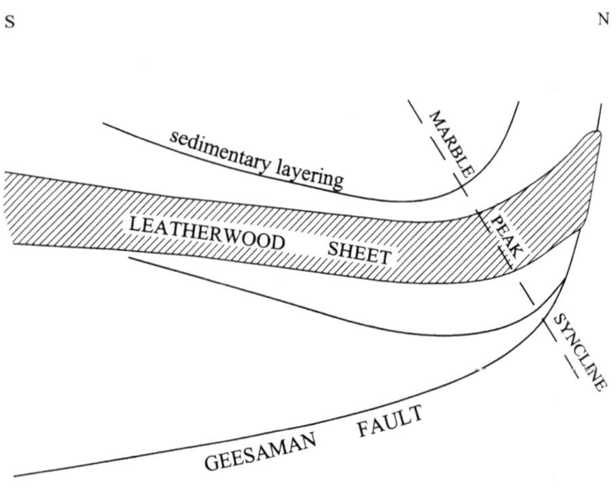

Figure 85 Diagrammatic cross-section of the Marble Peak syncline, showing relations to Leatherwood granodiorite and the Geesaman fault.

She found that folding dies out northward with metamorphic grade, but that the WNW lineation persists in association with top-to-east shear.

The recumbent folds are mostly evident only at outcrop scale (fig. 70). Locally, however, folding of this style has produced distinctive map patterns, such as the hour-glass shape of Mescal Limestone near Bear Wallow and an elongate syncline at Kellogg Mountain (pl. 1). These larger folds apparently have WNW trends, and their relation to elements of fabric is unclear. Locally, recumbent folding may be associated with thrust faults, as at Bear Wallow (pl. 1, loc. 21).

The folding predates intrusion of most Wilderness suite leucogranites, which truncate the folds (loc. 95) and clearly cut related fabrics (fig. 52). Spencer granite, the oldest of the Wilderness suite, both cuts and shows such fabrics. The association of Wilderness-related metamorphic grades with crenulated domains suggests that crenulation may be generally coeval with Wilderness intrusions.

The age of the main folding relative to older intrusions and the oldest faults is unclear, because these features mostly occupy different domains. However, Leatherwood granodiorite characteristically shows the foliation associated with the folds, especially in porphyritic Leatherwood sills (fig. 52), so we can presume that most folding is post-Leatherwood but pre-Wilderness.

On the northeastern face of Green Mountain, fabrics associated with folding are cut by Wilderness granite (fig. 15, loc. 4). Wilderness granite in turn shows mylonitic mid-Tertiary fabrics of the Windy Point zone about a kilometer to the west (pl. 1B). These age relations of the two fabrics are implied elsewhere in the study area but apparently displayed together only in such area.

Marble Peak syncline and related folds. Trending almost parallel to the Geesaman fault along the northern margin of Marble Peak is a syncline with a gently north-dipping southern limb, and a moderately south-dipping north limb (pl. 2B). It plunges about 10° to the ENE. The nearly concordant Leatherwood granodiorite at the base of the Paleozoic sequence is itself wrapped around this structure (fig. 85). This does not necessarily imply a post-Leatherwood fold, but the symmetry of both the bedding and the Leatherwood sill about the axial plane of the syncline does suggest that the Leatherwood is folded. Bedding is folded more tightly than the Leatherwood sheet—about 80° versus about 55° for Leatherwood. Probably Leatherwood was intruded after folding began, then was itself folded. The axial plane of the syncline dips northward, such that it intersects the Geesaman fault near the base of the Leatherwood sill.

For this reason, neither the base of the sill nor the Paleozoic section under it is deformed at the land surface.

In addition, Paleozoic sedimentary units at Marble Peak are intricately folded at outcrop scale, but these folds do not control the map pattern (pls. 1, 2B). Some of these folds are tight, and some recumbent (fig. 39). In some, Leatherwood dikes are folded (fig. 71), but in others, Leatherwood dikes cross folds without deflection, both consistent with relations for the Marble Peak syncline. Fold axes trend in several directions. I have not attempted to decipher the complex geometry of this folding, but broadly speaking the folding must have accompanied Leatherwood intrusion.

South of the Geesaman fault near the eastern margin of the study area, Janecke (1986, 1988) mapped some upright folds in Horquilla Limestone. These trend about N. 65° W. Janecke related them to Laramide movement on the Geesaman fault.

Pre-Leatherwood Faults

Some of the most important faults in this part of the study area apparently predate intrusion of Leatherwood granodiorite. These faults are described from north to south, because the Geesaman fault is the most important.

Geesaman fault. This structure forms the northern boundary of the area being considered here and is one of the most continuous structures of the study area. Janecke (1986, 1987) has described this fault in detail.

The Geesaman fault varies in attitude from east-west strike and steep southward dip at the west end, to northeast strike and shallow southeast dip at the east end. North of the fault are Precambrian rocks and Rice Peak porphyry, whereas south of it are Paleozoic strata and Leatherwood granodiorite. Carbonate-bearing rocks in and adjacent to the fault are commonly tectonites with subhorizontal lineation striking about parallel to the fault trace. Brittle gouge is also present in some exposures. Apparent dip-slip offset on the fault varies from less than 500 m (1600 ft) to about 1300 m (4250 ft), based on a stratigraphic separation diagram (fig. 86) extensively modified from that in Janecke (1987, her fig. 4). The irregularity of apparent throw is an indication that most motion was not dip-slip.

Where Leatherwood granodiorite forms the hanging or southern wall of the Geesaman fault, I find

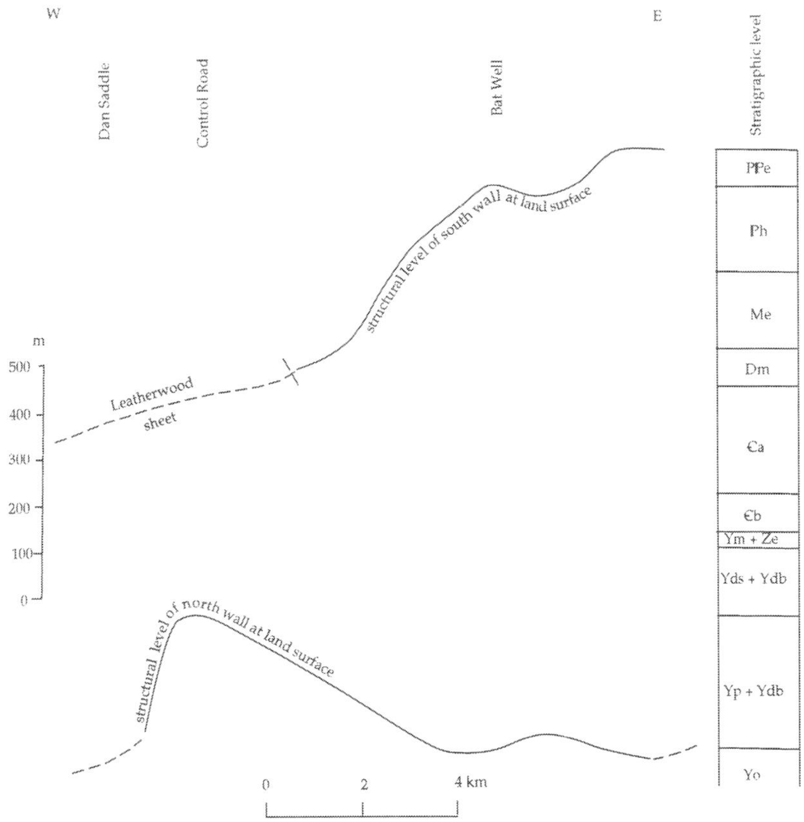

Figure 86 Stratigraphic separation diagram for the Geesaman fault, extensively modified from Janecke (1988, fig. 4). Unit designations from plate 1.

that it is most commonly intruded into the fault and chilled against the footwall. In such localities, all movement on the Geesaman must predate Leatherwood. This relation can best be seen on the Oracle Control road (fig. 87, loc. 64) and north of Marble Peak (locs. 110, 111). On the trail from Dan Saddle to Catalina Camp, Leatherwood intrudes mylonite of the Geesaman fault in a few places (loc. 112). East of the main Leatherwood body, porphyritic Leatherwood dikes cut the Geesaman fault and a splay of it (loc. 65). However, in the areas of Catalina Camp and the Corregedor mine, the Geesaman fault shows reactivation post-dating Leatherwood granodiorite, i.e. Leatherwood is locally sheared.

The Geesaman fault is intruded by Catalina granite at its western end (pl. 1); the Reef of Rock facies has intruded the fault for a short distance, and related felsites are intermittently found in the fault for over a kilometer (loc. 113). This relation is not surprising but has an interesting consequence, i.e. greisenization that is associated with the fault for at least 2 km, to Dan Saddle.

In Alder Canyon from Big Alder Spring to the eastern edge of the map area near Juan Spring, a subhorizontal fault separates a footwall of Pinal Schist from a hanging wall of gently dipping Paleozoic limestone (pl. 1). The Pinal thus forms a window, divided into two parts by later faulting. This fault I consider a splay of the Geesaman fault, which in its nearest exposure — about 1 km north and about 60 m higher — is dipping gently south. The fault plane contains both tectonite and gouge (like the Geesaman) but shows no lineation. Locally the dip of the fault is gently to the north. A Leatherwood dike cuts this fault (loc. 65), and wollastonite is locally present in Escabrosa Limestone just above it.

Relations on the Geesaman fault imply a relative age of the older facies of Rice Peak porphyry and Leatherwood granodiorite (fig. 49); Leatherwood intrudes the fault, whereas the older variety of Rice Peak forms the footwall and thus predates it. This relation suggests fault movement between about 79 and 66 Ma, i.e. Late Cretaceous. Rocks adjacent to the fault show strong foliation and(or) strong recrystallization. Some of the latter can be attributed to the heating effects of Leatherwood granodiorite.

Janecke (1988) interpreted the fault as having originated as a listric normal fault, concurrent with accumulation of the Bisbee Group in the Late Jurassic to Early Cretaceous, followed by Laramide(?) reactivation as a thrust fault with movement parallel to fault strike. The steep westerly portions would have functioned as a side ramp with left-lateral motion. Janecke's model has much to recommend it in the attitude, fabrics, and age relations of the fault, and is also consistent with the pre-Leatherwood age of faulting and the southward continuation of a splay of the Geesaman fault as a nearly horizontal plane to include the window in Alder Canyon. Pinal basement in this window is consistent with her composite nature of the Geesaman fault.

On the other hand there are problems with the Janecke (1988) model. Her plot of elevation versus fault dip produced a smooth curve from which she inferred a concave-up fault surface and thus a listric ancestry. However, variations in the trend of the fault are not consistent with her model. In addition, fault dips as corrected for mid-Tertiary tilting, i.e. north dips near Dan Saddle, are not suggestive of listric faulting. The association of the Geesaman fault with probable Bisbee Group is consistent with the Janecke model, but cannot be considered as strongly supportive of it in view of the uncertainty of the assignment as Bisbee Group, discussed with the description of that unit.

The possibility that the Geesaman fault is simply a deformed plane needs to be addressed. The geometry of the Marble Peak syncline (pl. 1, fig. 85), suggests that the Geesaman fault must be folded, because the Leatherwood granodiorite, which intrudes the fault, is itself folded. All that is required to produce the observed shape of the Geesaman fault, then, is for the axis of this fold to become tangential to the trace of the Geesaman fault toward the east. The Geesaman fault would dip gently southward near and south of the axis. The eastward plunge of the fold and the near-parallelism of fold and fault then explain the flaring fault trace toward the east.

West of Marble Peak, relations are obscured by Leatherwood intrusion, but the intermittent yet persistent association of steeply dipping Abrigo Formation

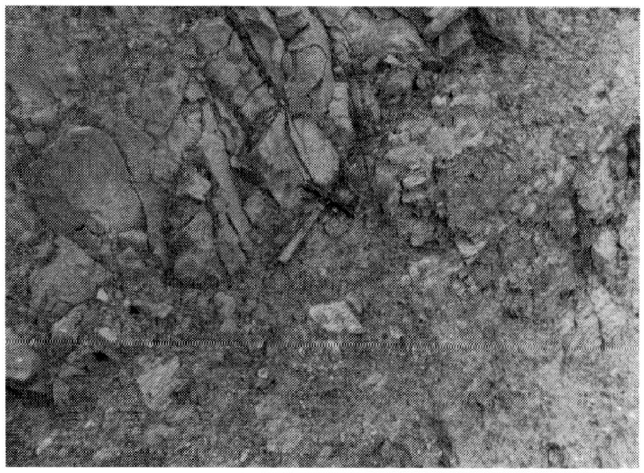

Figure 87 Leatherwood granodiorite intrusive into the Apache Group footwall rocks along the Geesaman fault, Oracle Control road (loc. 64). Contact above hammer handle.

with the steep Geesaman fault all the way to the Catalina granite suggests that the relation at the west end of Marble Peak continues; that is, the fault remains on the north limb of the fold and so remains steep.

Thus the Geesaman fault probably originated as a steep planar feature. In conjunction with fabrics in the fault plane, parasitic folds, and stratigraphic separation, this form suggests that the Geesaman fault could have originated as a left-oblique normal fault. If so, the apparent offsets of several Precambrian features shown on plate 1 would provide the magnitude of the strike-slip component. Among these are the belt of Pinal basement, the sandy middle member of Pioneer Formation, the coarse pyroclastic facies of Pioneer Formation, the lacustrine facies of Dripping Spring Quartzite, and the diamictite facies of Campo Bonito Formation. The suggested magnitude of Laramide left-lateral strike-slip component of faulting is 7 km, a magnitude consistent with the absence of older-variety of Rice Peak porphyry south of the fault.

Faults at Hartman-Homestake and Leatherwood mines. Some of the skarn deposits of the Marble Peak district (pls. 1, 2B), along intrusive contacts of Leatherwood granodiorite with Paleozoic limestone, are localized along steep faults striking about N. 55 to 75° W. Details of the geology are described under mineral resources. These faults are pre-Leatherwood; they do not extend into immediately adjacent Leatherwood, and Leatherwood dikes are found along them. The movement sense is generally south-side-up from 10 to 20 m. For a similar probably pre-Leatherwood fault at the Daily mine (as shown on the topographic map), marked only by a step in the Leatherwood contact, the movement sense is unclear. These faults are about parallel in strike but not dip to the nearby Geesaman fault of about the same age.

Red Ridge fault. On the lower part of Red Ridge, a steep fault striking about N. 45° W. repeats Pioneer Formation and juxtaposes it against Dripping Spring Quartzite. This fault appears to be cut by Leatherwood granodiorite and Catalina granite (pl. 1). Offset is down on the south, possibly as little as 100 m (330 ft).

Edgar and Araster thrust faults. Along the upper slopes in Alder and Edgar drainages, Bykerk-Kauffman (1990; written commun., 1991) mapped a north-dipping fault that separates Precambrian rocks including Pinal Schist basement to the north from Abrigo Formation to the south. She called it the Edgar thrust fault, and documented a pre-Leatherwood age. It extends eastward beyond the study area and has a total strike length of over 25 km. Bykerk-Kauffman found elsewhere that the fault cuts up section to the southwest, suggesting upper-plate transport in that direction.

The Edgar thrust dips about 20° to the north. Intense rodding trends almost parallel to fault strike (pl. 1), i.e. WNW. A mappable body of porphyritic Leatherwood cuts the fault. Stratigraphic separation is locally as much as about 350 m, but the fault intersects bedding at a low angle, so that offset on the fault plane could be significantly greater.

In the eastern part of the Edgar fault as mapped, I find that the parallel Araster fault (for Araster *sic* Spring) assumes most of the motion, juxtaposing Abrigo Formation to the south against various parts of the Apache Group. In this interval, the Edgar fault mostly juxtaposes different parts of the Apache Group. Thus the Edgar trend can be considered a family of thrust faults with an aggregate stratigraphic repetition of about 350 m.

West of the intrusive contact of the main Leatherwood sheet, the relations characteristic of the Edgar thrust continue as a relation between the Mt. Lemmon and Marble Peak areas, separated by the Leatherwood intrusive sheet (pl. 1). The Edgar fault reappears on Red Ridge between Pioneer Formation pyroclastics to the north and Abrigo Formation and the Mt. Lemmon unit to the south, with a small Leatherwood dike in the fault plane. Another intruded segment of the Edgar thrust probably continues to the Catalina granite.

Canada del Oro headwater area. In this area, two faults may predate Leatherwood granodiorite, but most of their courses run through Leatherwood, where offsets can not be detected and traces are hard to find. The first apparently trends down the east fork of Canada del Oro and is responsible for the differing stratigraphies of Oracle Ridge and Red Ridge (pl. 1). The apparent offset between Marble Peak and Red Ridge is east-side-down, whereas the apparent offset of the top of the Leatherwood sheet is in the opposite sense, suggesting pre-Leatherwood motion (fig. 88). Where this fault cuts Paleozoic rocks at the south end, the west side is downthrown, probably about 150 m (500 ft). The Leatherwood contact is apparently offset in the same sense, but less than the offset of sedimentary rocks. Where this fault crosses the Geesaman fault to the north, the apparent offset could be intrusive, as the Leatherwood intrudes the Geesaman fault.

A second fault runs east-west in Leatherwood at the northern foot of Red Ridge (pl. 1). In the Catalina Camp area, Abrigo Formation is persistently plastered against the Geesaman fault, whereas Apache Group occupies slopes across the valley to the south. There must be an up-to-south fault, probably pre-Leatherwood, as it is apparently cut by the east fork fault before reaching Marble Peak.

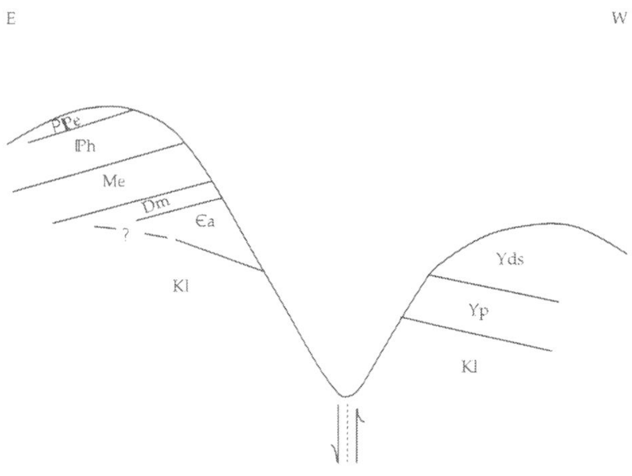

Figure 88 Cross-sectional diagram across east fork of Canada del Oro between Marble Peak and Red Ridge of relations that suggest a pre-Leatherwood fault. Units from plate 1.

Post-Leatherwood to Syn-Wilderness Faults

Bear Wallow thrust. In the Bear Wallow area, a deformed subhorizontal fault locally places Dripping Spring Quartzite over Mescal Limestone (fig. 89). This structure, the Bear Wallow thrust fault, is cut by several younger faults, but has been traced eastward to the area of Turkey Flat (pl. 1). The thrust is locally parallel to bedding, but it locally cuts bedding in overlying quartzite at 20 to 30°. Fault gouge is about 1 to 2 m thick. The fault is approximately parallel to axial planes of recumbent isoclinal folds in adjacent rocks; the fold axes trend about N. 45° E. However, Z-shaped folds in gouge itself (fig. 89B, loc. 21) suggest movement of the upper plate toward N. 80° W. Leatherwood granodiorite is cut by the thrust in poor exposures in the northeastern part of Bear Wallow. Some Wilderness suite pegmatites cut the thrust, but others are cut by it (loc. 21).

Kellogg and Knagge faults. North of the range crest in the Mt. Bigelow-Kellogg Peak area are two faults that dip southward at about 40°. They are reverse or thrust faults, as they are south-side-up (pl. 1). The Kellogg fault separates synclinally folded Pioneer Formation to the south from south-dipping Dripping Spring Quartzite to the north (fig. 90). Stratigraphic offset is about 100 m. Foliation as well as bedding has been rotated. The Knagge fault separates Pioneer Formation to the south from Paleozoic rocks on the north, and offset is probably over 200 m. Its fault plane is approximately parallel to south-dipping bedding and foliation on both sides, so true offset cannot be specified and could be large. Both faults are intruded by the Knagge pluton (loc. 114), but the Kellogg fault post-dates the development of foliation in metasedimentary wall-rocks, and probably therefore is post-Leatherwood. These two faults are separated from the Bear Wallow thrust by the younger Sphinx fault, but the Kellogg fault could be correlative with the Bear Wallow thrust. Some other faults probably of the same generation are shown by Waag (1968).

Marshall Gulch area. The present site of a pegmatite in Marshall Gulch must have been a fault separating Dripping Spring Quartzite to the south from Pioneer Formation to the north (pl. 1, loc. 99). Offset was probably about 100 m, down on the south. The fault plane itself has not been seen; no extension east of Sabino Creek could be located. Nearby, several other faults engulfed by pegmatite offset the Oracle-Scanlan unconformity, as along the Sunset Trail (loc. 115) where about 60 m of offset in the opposite sense has occurred. These faults form an irregular array (pl. 1).

"Dead Pine shear zone." Guerin and others (1990) proposed a "Dead Pine shear zone" of Paleocene age near the Catalina range crest. Associated lineations

Figure 89 Bear Wallow thrust along the Mt. Lemmon highway (loc. 21). Dripping Spring Quartzite is thrust over Mescal Limestone. A. Outcrop view; B. Kink folds in gouge.

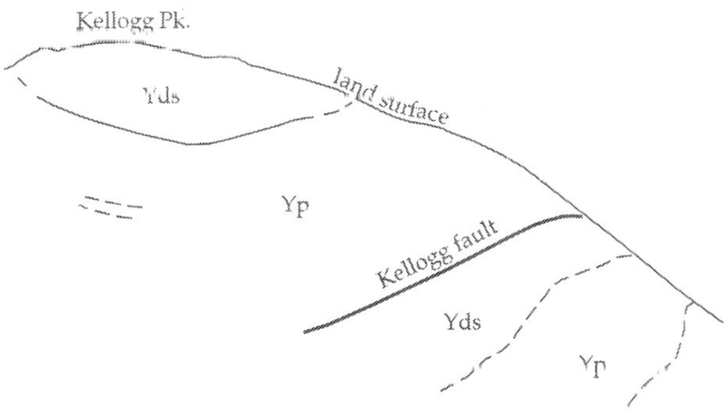

Figure 90 Kellogg fault from Knagge Trail looking NW, showing relation to syncline on Kellogg Peak; from a mosaic of photos taken at locality 114. Yds, Dripping Spring Quartzite; Yp, Pioneer Formation.

trend east-west. An unpublished manuscript by Guerin and others suggests that the host rocks are Leatherwood granodiorite and Apache Group in tight to recumbent sheath folds. The apparent correlation is with Laramide structures described above. However, maps and cross sections in these papers show the Dead Pine shear zone along the intrusive contact with Eocene leucogranites of the Wilderness suite. This contact is sheared locally, but the shearing cannot be Paleocene. More probably it is related to the Windy Point mylonite zone. I conclude that the Dead Pine shear zone of Guerin and others is a mixture of phenomena of different ages and origins.

Younger Faults

Most of the younger faults of this structural block are shown on Plate 1 but not described here. A few are of special interest.

Soldier Camp fault set. North-south steep faults through the Soldier Camp-Bear Wallow area number at least three, and occupy an interval of about 400 m width (pl. 1). They apparently cut the Bear Wallow thrust fault. Toward the northern end of the fault set, the western fault juxtaposes Dripping Spring Quartzite against Pioneer Formation, and the eastern fault may separate Paleozoic rocks of the Butterfly Peak area on the east from Dripping Spring Quartzite on the west, for a total offset of at least 100 m, east side down. Toward the south, the apparent offset of the Spencer leucogranite on the eastern fault is opposite to that of older supracrustal rocks, suggesting precursor pre-Spencer faults. The western strand extends farthest south, offsetting slightly the Wilderness-Spencer screen, but is separated from the Bear Canyon fault of the forerange (pl. 1) by a zone of dense north-south jointing without offset.

Westfall fault. An apparently steep east-west fault through the Westfall Knob area is inferred to separate Paleozoic rocks to the north from older rocks to the south. The fault also truncates numerous Leatherwood sills. Apparent offset is 100 m or more down on the north. Plate 1 shows the fault in two segments, separated by the younger Sphinx fault.

Sphinx fault. North from Palisade Spring, the steep Sphinx fault extends past a crag known as Sphinx Rock and then Westfall Knob, to disappear northeastward under a large colluvial fan (pl. 1). It is upthrown on the east; the amount of offset is difficult to measure because of the abundance of porphyritic Leatherwood sills in Paleozoic rocks. Fault strike is about parallel to that of the Soldier Canyon fault set, but movement is opposite, so that the area between is a graben. Correlation across the Sphinx fault of 1) the Kellogg fault with the Bear Wallow thrust, 2) the syncline of the Kellogg Peak area with that at Turkey Flat, and 3) the two segments of the Westfall Knob fault is possible, but only if the Westfall Knob fault and the axial plane of the syncline dip southward, and if movement on the Sphinx fault has a right-lateral component.

Structural Restoration

The restoration of features in this study-area segment is tentative, because of the uncertain and/or overlapping relative ages of some of the more important features. Such complications led to the approach taken in figure 47 for "bulk restoration" for the situation late in the Laramide time interval. The question of restoring the deformed Geesaman fault is deferred to the next section, when both walls will have been described.

Tilting and late structures. The area being considered here has no direct evidence of mid-Tertiary tilting, because of the lack of mid-Tertiary units. However, we can presume that it was so tilted because formerly horizontal beds of Precambrian through Early(?) Cretaceous age dip gently northeast in most of the area, and this tilt domain passes both northward and southward into areas where the same tilts are known to be mid-Tertiary. The axis of tilt is uncertain, but is about N. 30° W. north of the Geesaman fault and N. 60° W. south to the range crest.

Restoration by back-tilting about 20° has the following effects (fig. 91): 1) The Geesaman fault has a steep northward dip at its west end, warping around eastward to a steep, then moderate, southward dip. 2) The Leatherwood sheet dips gently southward except at Marble Peak. 3) The Marble Peak syncline reverses its plunge and becomes a monocline, gently dipping toward the south but steep on the north, with a moderately north-dipping axial plane. 4) The Edgar and Araster thrust faults are about horizontal. 5) The Bear Wallow thrust is almost horizontal in some segments and gently south-dipping in others, but the otherwise similar Kellogg and Knagge faults dip steeply south. 6) Bedding and sills throughout the area become nearly horizontal, with some domains of southward dip near the range crest that must have resulted from Laramide deformation. The attitude of the steep northeast-striking margin of the Leatherwood sheet (fig. 45) is virtually unaffected, as are all steep northeast-striking faults. The WNW-trending folds and fabrics are little-affected except in dip. Restoration of the young faults has no regional effect, but slightly elevates a block from Butterfly Peak to Novio Spring, i.e. between the Soldier Canyon and Sphinx faults.

Tertiary intrusions. In the Barnum Rock-Knagge Trail-Green Mountain area, the juxtaposition between Precambrian and Paleozoic rocks produced by removal of the Knagge pluton suggests that the Knagge fault once extended far to the southeast, for a total length of at least 4.2 km.

Removal of Wilderness suite sills allows the reconstruction of original basement-cover relations along the crest of the range (fig. 91). In some areas, screens fit perfectly into slots in the main sequence, as between Spencer Peak and Palisade Spring where a long screen of Pioneer Formation matches an interval of missing Pioneer at the base of the supracrustal sequence (pl. 1). In the Kellogg Mountain and upper Sabino Canyon areas, removal of the Wilderness granite sill leaves Scanlan Conglomerate Member of Pioneer Formation resting on Oracle Granite or Pinal Schist. In such areas, Wilderness sills have clearly split open an older unconformity or other contacts with minimal disruption even where these sills are thousands of feet thick.

In the Marshall Saddle area, removal of Wilderness granite leaves a stratigraphic question. Apache Group is missing between Bolsa Quartzite and Lemmon Rock pegmatite-aplite. Nearby, Apache Group intervenes in condensed sequences, as at Lemmon Rock where all but Pioneer Formation is missing. Therefore

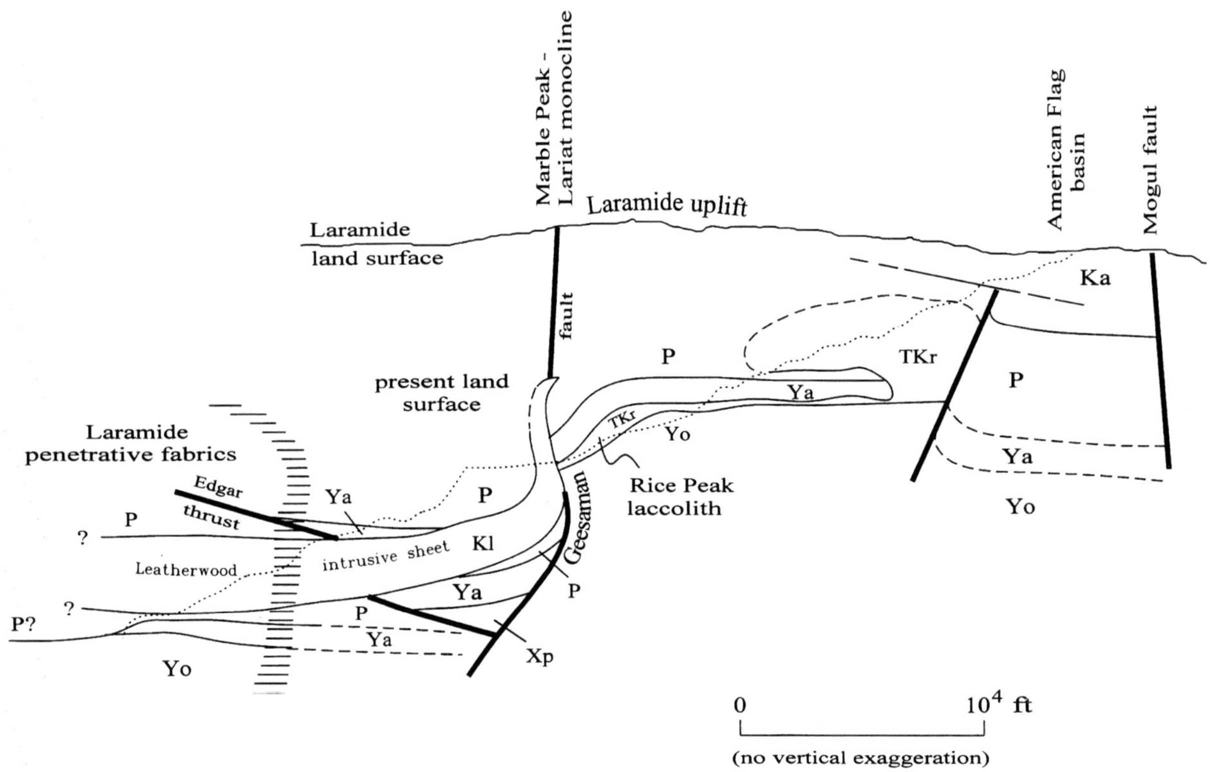

Figure 91 Cross-section restored for mid-Tertiary tilts and Wilderness -suite intrusions, i.e. showing Laramide situation, modeled on section B-B' of plate 1. Note that further removal of Laramide intrusions restores the older supracrustal sequence. Laramide penetrative deformation minimized because fabrics are at a high angle to illustration plane.

near Marshall Saddle, it is difficult to say whether Apache Group was displaced by Wilderness or eroded before Bolsa deposition.

Removal of Wilderness sills also reveals a previous stage of Leatherwood geometry. The distribution of inclusions (pl. 1B) and a few contact relations suggest that the Leatherwood sheet had intersected the unconformity on Oracle Granite in the headwaters of Sabino Creek near Marshall Gulch. A subsidiary Leatherwood sill was emplaced in Pioneer Formation near Lemmon Rock. As we have seen, deformed scraps of Leatherwood are recognizable to the south as the sill continued to cut downsection.

Deformation along the range crest. Restoration of range-crest deformation broadens this part of the study area in an east-west direction, i.e. not in the plane of figure 91. Restoration of the Bear Wallow thrust probably contributes about 700 m in a WNW-ESE direction, in addition to that required by associated folds. If the Knagge and Kellogg faults have the same age and movement direction, restoration requires about another kilometer of movement. Sheath-fold geometric relations of fabrics suggest translations of at least the width of such domains, here more than a kilometer in about the same direction. The total translation is thus over two kilometers. Restoration of the sheath fold in top-to-west mode as with the Bear Wallow thrust straightens the bend in the boundary of Pinal and Oracle basement types (pl. 1B).

At this point in the restoration we are dealing with a complexly faulted and intruded supracrustal sequence that commonly dips gently south unless such dip is removed along with tight folds. This dip might be a clue to broader-scale Laramide deformation.

Leatherwood granodiorite. Removal of Leatherwood restores the supracrustal section almost perfectly on the southeastern margin of the Leatherwood sheet, but suggests a number of faults to the west that are probably pre-Leatherwood (fig. 88). Intrusion of Leatherwood across the Geesaman fault allows land-surface data from north of the fault to be extrapolated south of the fault for the time of intrusion, suggesting depths of burial ranging from about 2 km near the Geesaman fault to about 4.1 to 5.5 km at the range crest (fig. 47). Post-Leatherwood movement due to folding or faulting modifies this extrapolation, but the paleo-vertical components of such movement were modest except for the Marble Peak syncline (next section).

Pre-Leatherwood faulting. Restoration of pre-Leatherwood faulting and tilting leads to a fully restored horizontal supracrustal sequence of Proterozoic and Paleozoic rocks about 1.4 km thick, resting on Oracle Granite. The faults restored include two thrusts (Edgar and Araster), a left-oblique normal fault (Geesaman), and steep faults, both north-side-up (Red Ridge) and south-side-up (Hartman-Homestake, Daily). Such disparate fault styles probably formed at different times.

Except for the Geesaman fault, for which restoration is described in the next section, the Edgar thrust set probably forms the most important structures. Restoration of these thrusts alone decreases the Laramide structural thickness at the range crest by over 600 m, i.e. to about 3.5 - 4.9 km or less. The amount of translation implied is at least a kilometer. Because Apache Group and Paleozoic rocks of the upper plates appear to have ridden about a kilometer from the east, restoration requires broadening this part of the study area by translating the upper-plate to the east. The apparent southern end of the lacustrine facies of Dripping Spring Quartzite (pl. 1B) is probably due to this translation. The sense of movement of the Edgar thrust set implies that its Laramide strike was approximately north-south.

Summary of Laramide restoration. In aggregate, restoration of Laramide features implies that east-west crustal shortening of 3 km or more occurred south of the Geesaman fault. This magnitude of shortening is modest, but is greater than the stretching implied by mid-Tertiary mylonitization in the study area.

The "snapshot" of burial depths for the time of Leatherwood intrusion, from 2 km at the Geesaman fault to over 4 km at the range crest (fig. 47), can be decreased at the range crest by about 350 m prior to thrusting on the Edgar and Araster faults. After intrusion, reconstruction requires a further decrease of perhaps as much as 150 m on the Bear Wallow thrust. Thus the total crustal thickening due to thrusting at the range crest, about 0.5 km, is quite modest. It cannot compare with crustal thickening due to local intrusions, at least 4 km.

GEESAMAN TO MOGUL FAULT

This structural interval is the domain of Rice Peak porphyry and the American Flag Formation. The most conspicuous structures (pl. 1) are the Lariat anticline and Oro syncline, and faults like the Limb fault toward the west. Less obvious but important structures are the reverse faults bounding the American Flag basin and the Stratton detachment fault. The interval is bounded by the Geesaman and Mogul faults that apparently were both Laramide left-oblique faults with large displacements. The interval thus has a history independent of adjacent blocks before the Laramide, and the same history subsequently. Metamorphism is low-grade in this interval, but foliation is strong in the more susceptible rocks. General severity of fracturing increases northward.

Structures are described in order of decreasing age. The Geesaman fault and Marble Peak syncline have been described above but will be reconstructed in this section.

Laramide Structures

Faults synchronous with older Rice Peak porphyry. Several northeast-striking steep faults between Stratton and Geesaman washes (pl. 1) are intruded by Rice Peak porphyry dikes (apparently of the older variety, loc. 12). Intruded extensions of these faults can be mapped through larger bodies of Rice Peak porphyry by apparent offsets between large included blocks of sedimentary rock. The faults traverse structural sections from Oracle Granite through Escabrosa Limestone. The vent facies of Rice Peak porphyry is probably along one of these faults (pl. 1). Movement sense is not consistent among the faults.

Janecke (1986, fig. 30) found that these faults can best be restored if the Rice Peak porphyry bodies are removed; then all the older rock bodies including diabase can be brought into alignment by small movements on the faults. Some of her few remaining misalignments can be explained by her mapping of pyroclastics in the Pioneer Formation as Rice Peak porphyry. She suggested that faulting and Rice Peak intrusion are related and synchronous, and referred to the faults as release structures formed by intrusion. My restoration of these structures (below) starts where Janecke left off.

Lariat anticline and Oro syncline. Two major structures that may possibly be as old as Rice Peak porphyry are the Lariat anticline and the probably related Oro syncline, which control much of the map pattern in the area being considered (pl. 1). The Lariat anticline can be traced for about 10 km, and has a core of Oracle Granite, flanked by Apache Group (fig. 14). The anticline may continue northward to the Mogul fault, but its eastern limb is intruded and its western limb faulted in this northern region. The anticline is intruded by Rice Peak porphyry and variably faulted throughout its length, somewhat obscuring regional relations. However, the anticlinal nature is clearer on plate 1 than on previous maps, because the Pidgeon Tank and Bat Hill faults (of Janecke, 1986; see also Creasey and Theodore, 1975) are shown as minor structures thereon. I have found that these faults as mapped are between Oracle Granite and Scanlan Conglomerate Member in all but one area. Both rocks are severely strained, but there is no evidence of offset and no other evidence of faulting (locs. 116, 117). Thus this contact is an unconformity, and the region of the Pidgeon Tank and Bat Hill faults is simply the south limb of the Lariat anticline. An approximate angle between the limbs of the anticline can be given as 45 to 70°.

The Oro syncline to the west of the Lariat anticline (pl. 1) occupies much of upper Canada del Oro. The two structures were probably originally adjacent, but are now separated by the well-named Limb fault of Suemnicht (1977) and related faults (pl. 1). The syncline is also truncated on the south by the Limb fault. The Oro syncline is shown by Wallace (1955), Suemnicht (1977), and on an unpublished field sheet of S. C. Creasey. It has Mescal Limestone in the core for much of its length (loc. 22), but the core locally consists of Dripping Spring Quartzite or rocks as young as Abrigo Formation, due to severe cross-faulting. Too little of the eastern limb is preserved to give an angle between limbs.

The Oro syncline is truncated on the north by early movement on the Mogul fault, and the Lariat anticline is truncated on the south by the Geesaman fault, which juxtaposes the core of the anticline on the north side against a north-dipping section on the south side. To the west and south, Catalina granite has intruded one limb of the Oro syncline.

Evidence constraining the age of the Lariat anticline include truncation by the pre-Leatherwood Geesaman fault, and restriction of Rice Peak porphyry (mostly the old variety) to the northeast limb. Thus the anticline must be a relatively old Laramide feature. The Oro syncline is probably the same age, but evidence is limited to truncation by Catalina granite.

Fault of Mary West mine. A steep north-trending fault is visible in the workings of the Mary West mine (loc. 118). The map pattern (pl. 1) suggests that this fault may be synchronous with intrusion of Rice Peak porphyry. The fault plane is intermittently mineralized from its intersection with porphyry for about a kilometer to the north.

American Flag reverse faults and related folds. Description of the American Flag Formation has shown that it accumulated in a time period that spanned faulting along the margins of its basin and intrusion of Rice Peak porphyry (younger variety). The faults now dip gently southwest and locally place Escabrosa Limestone above the lower member of American Flag Formation, commonly overturned (pl. 2A, locs. 46, 47, 54). Appreciable offset on these faults is suggested by tectonic inclusions of other units in foliated gouge (loc.

119). Steeper tear faults join segments of the more gently dipping faults.

Provenance of cobbles in the middle member, which overlaps the faults, shows that the paleo-topography on the upper plate exposed Oracle Granite and Apache Group. Probably the reverse faults record shortening in a WSW-ENE direction, consonant with the direction of tectonic transport given by rock fabrics (Bykerk-Kauffman, 1990) to the south.

Synclines in the Rice Peak and Apache Peak areas and an anticline in the Peppersauce Wash area deform the northeast-dipping limb of the Lariat anticline into an open S-shape (pl. 1; shown best by Pioneer Formation). American Flag Formation may have accumulated or been preferentially preserved in the synclines (pl. 1); if so, these folds were forming concurrently. The younger variety of Rice Peak porphyry is also most common in the southern syncline. These folds are cut by the Limb and related faults and by the early phase of the Mogul fault.

Proto-Mogul and related faults. The Mogul fault, which forms the northern boundary of this area, has had a complex history. Segments showing older motion have been mineralized, intruded, and offset by younger faults (pl. 2C). Younger motion abandoned these older offset segments for a plane that is within Oracle Granite for much of its length (pl. 1). Henceforth, the older offset segments are collectively called the proto-Mogul fault. They are easy to map because they juxtapose Oracle Granite to the north against supracrustal Proterozoic through probable Cretaceous rocks to the south.

The proto-Mogul fault is vertical to steeply south-dipping. Gouge is generally minor (loc. 120) but Oracle Granite is locally represented by chloritic breccia within a few meters of the fault. Sense of movement is unclear from fabrics in the fault zone. Between the Campo Bonito district and the 3C Ranch (loc. 121), the fault plane is overprinted by later movement, as discussed below.

The proto-Mogul fault is best known in the Campo Bonito mining district (pl. 2C). The fault plane itself is best exposed at the mouth of the Cody tunnel (loc. 122), where it is vertical and separates protoclastic Oracle Granite from Dripping Spring Quartzite intruded by diabase. Across the gully to the west, however, the south wall consists of Escabrosa Limestone, and the fault has been offset to the south. Intruded along the offsetting fault is an east-dipping porphyry dike, now weathered but probably younger Rice Peak porphyry (see loc. 122).

Approximately parallel to and south of the proto-Mogul fault are more steep faults (pls. 1, 2C); they are probably related to the proto-Mogul, as they are offset in a similar way. The proto-Mogul faults and these parallel faults to the south locally enclose domains of rock oriented parallel to them, probably due to rotation. Creasey (1967) noted that these domains are intricately faulted internally. Locally, foliation of Pinal Schist is dragged into conformity with these structures also.

The proto-Mogul fault cuts the lower member of the American Flag Formation but is cut by probable younger variety of Rice Peak porphyry. Thus movement must be roughly synchronous with accumulation of the middle and(or) upper members of the American Flag Formation (fig. 49). Toward the west, the Limb and related faults are among the cross-faults that cut the proto-Mogul fault.

The apparent throw is generally down on the south, 700 m or more in some places. However, where the fault juxtaposes Oracle Granite against Pinal Schist, dip-slip movement is unclear and may be minimal. Left-lateral strike-slip movement is suggested by offset of the steep Oracle-Pinal contact. Thus the proto-Mogul is probably a left-oblique normal fault.

Banerjee (1957) mapped drag folds outlined by inclusion trains in Oracle Granite just north of the proto-Mogul fault (but west of the map area). My stereographic plots of Banerjee's data suggest drag-fold axes trending about S. 56° W., plunging 60° SW. However, nearest the study area the axis plunges about 74° SW. If the slip vector is orthogonal to this axis, it plunges 16 to 30° to the southeast, supporting the left-oblique nature of the fault. Apparently movement of the north side was up and toward the northwest. If the minimum dip-slip component is 700 m, the corresponding minimum strike-slip component is 1.1 to 2.4 km. The apparent offset of the Oracle-Pinal contact is over 7 km, and this may be the best measure of offset because its dip is at a high angle to offset.

Limb and related faults. Three faults, all steep and striking roughly north-south, trend along the eastern slopes of Canada del Oro (pl. 1). Suemnicht (1977) mapped the westernmost and named it the Limb fault. It dips steeply westward (loc. 123) and separates rocks near the core of the Oro syncline, as young as Bolsa Quartzite, from Pinal Schist and Oracle Granite probably in the core of the Lariat anticline. Thus its throw is at least 350 m down to the west at such localities. The offset of the middle fault of the set is unknown. The eastern fault, striking east of north, is also down to the west; at various places it juxtaposes Pioneer Formation to the west against Oracle Granite to the east (loc. 6), and Dripping Spring Quartzite to the west against Pioneer Formation to the east (loc. 124). Along most of its length, relations are obscured by Rice Peak porphyry (mostly older variety), which is cut by the fault (loc. 6) but commonly present on both walls (pl. 1). At the north end, separation of the Pioneer-Dripping Spring contact is about 40 m (120 ft). The middle fault probably does not extend to the proto-Mogul fault, but the other two both offset the proto-Mogul in a left-lateral sense. The Limb fault is cut by the intrusive contact of Catalina granite (pl. 1).

Tertiary Structures

Stratton detachment fault. In the valley of Stratton Wash are poor exposures (locs. 125, 126) of a structure separating the Stratton unit (Oligocene) to the northeast from Apache Group and Rice Peak porphyry (older variety) forming the footwall to the southwest (pl. 1). Locally, Oracle Granite (not mapped) and Paleozoic rocks form the footwall. Cataclasis is common in footwall rocks but no microbreccia ledge is present. The attitude of the fault varies from subhorizontal, to northeast or southeast dips up to about 45°. Dips in the Stratton unit are regular and steep to the northeast. That is, the detachment fault cuts the bedding of upper-plate rocks of synextensional origin, judging from the immature, proximal debris. Probably, the Stratton fault is an extension of the San Pedro detachment fault of Dickinson (1991) which is itself an apparent continuation of the Catalina detachment fault. Topologically, these faults together suggest the continuation over the Catalinas of a Catalina-Stratton detachment system.

Basin-range faulting. A steep basin-range fault cuts the Geesaman fault and the Stratton detachment fault at their eastern ends (pl. 1). East of this structure is basin fill of the San Pedro basin. A similar fault separates American Flag Formation from basin fill (pl. 2A). Due to gravel-capped pediment surfaces, such basin-range faults are difficult to map in this area, but the alignment of steep contacts between bedrock and basin-fill units (pl. 1) suggests that basin-range faults may commonly separate the Geesaman-Mogul bedrock block from the San Pedro basin. South of Davis Spring, some of these structures probably enter the main range.

Neo-Mogul fault. Within Oracle Granite north of the proto-Mogul fault are shear zones (locs. 127–129) that represent westward projections of offset proto-Mogul segments. There are probably more of them than are shown (pl. 1), because they are hard to map. These faults, together with the plane shared with the proto-Mogul from the Campo Bonito district eastward to 3C Ranch, are a reactivated Mogul fault zone, and will be referred to as the neo-Mogul fault.

East of the 3C Ranch, two newly discovered splays of the neo-Mogul fault are present (pl. 1). These splays separate alaskite related to Oracle Granite on the north, from the Stratton unit between the splays, then Quiburis Formation without alaskite clasts on the south (locs. 59, 130). Thus movement on the neo-Mogul fault postdates the Stratton unit, Stratton detachment faulting, and some Quiburis Formation.

Fault planes on the neo-Mogul fault are vertical to steeply south-dipping. Where they are within Oracle Granite they show quartz-sericitic gouge with quartz or quartz-chlorite gash veins in cataclastic granite wallrock (locs. 127–129).

Where the neo-Mogul juxtaposes granitic rocks and Tertiary deposits, both are shattered, with vague rodding both gently and steeply plunging (locs. 59, 130). The splay between Stratton unit and Quiburis Formation is almost vertical, whereas that between alaskite and Stratton dips steeply south. The Stratton unit contains quartz-chlorite veins there.

As with the proto-Mogul fault, the apparent dip-slip motion is down to the south. The magnitude of dip-slip motion is not known; a minimum figure is given by the vertical relief across each splay of the fault in Peppersauce Wash, for a total of about 70 m (loc. 59). The lack of cataclasis attributable to the footwall of the Stratton detachment fault adds approximately 10 m to this minimum figure, and dip-slip movement of over 100 m is probable. Some left-lateral motion is also indicated by offset diabase dikes in Oracle Granite on the northern flank of the zone (Banerjee, 1957). Left-lateral strike-slip offset of about one kilometer would explain the lack of alaskite clasts in Quiburis Formation facing the fault, assuming that some deposition was from the rising block to the northeast. Alternatively, the neo-Mogul may completely postdate Quiburis deposition. The alaskite evidence thus indicates that the neo-Mogul fault is either a left-oblique normal fault or a very young fault.

The neo-Mogul faults hide some segments of the Stratton detachment faults, i.e., at exposed levels, the south wall of this fault system is partly in upper plate and partly in lower plate rocks of the main detachment system of the Catalinas. As we will see, the north wall at exposed levels is above one detachment system and below and/or beside another. Thus, the neo-Mogul fault makes detachment-system correlation difficult.

Structural Restoration

Prior to movements on the proto-Mogul and Geesaman faults, features in this area can be restored independently of adjacent areas. Following juxtaposition on those faults, restoration of one area has implication for the others.

Mid- to late-Tertiary features. Restoration of the youngest movement on the Mogul fault system involves motion of the northern (Oracle) block down at least 80 m and to the east possibly more than a kilometer. Restoration thus increases the topographic paleo-relief on the fault scarp; this may have some importance in the formation of rock-avalanche deposits in the San Manuel Formation.

Mid-Tertiary extension in this part of the study area probably consisted of detachment faulting with coeval tilting of the upper plate, followed by further northeast tilting of all the rocks as a consequence of core-complex uplift (growth of the Forerange anticline). Restoration thus requires back-tilting of the entire block, followed by restoration of the Stratton detachment fault and the tilting of the Stratton unit.

Restoration (fig. 91) tilts the sequence up to the ENE from 15° to 30°. This makes the older parts of the Mogul fault system (the proto-Mogul) vertical to steeply northeast-dipping, i.e. a left-oblique reverse or strike-slip fault. American Flag Formation has a northeast dip of 10 to 30° (except along its southeastern margin). The reverse faults associated with American Flag Formation have a steep to moderate southwest dip. The Stratton detachment fault probably averages about horizontal; the Stratton unit dips about 40° into the fault plane. Proterozoic and Paleozoic supracrustal rocks are subhorizontal to gently east-dipping over most of the region but complexly faulted, and gently folded along subhorizontal axes into the northeast-trending synclines at Rice Peak and Apache Peak, and the anticline at Peppersauce Wash. The Lariat anticline becomes asymmetric, with a long gentle northeast limb and a short steep southern limb. Plunge is reversed toward its eastern end to become gentle to the west. Similarly, the Oro syncline becomes asymmetric with a gentle western limb. The intervening Limb fault and related faults become vertical. Rice Peak porphyry becomes a subhorizontal sill or laccolith with derivative bodies configured in cross section as shown in figure 48 (except for faulting). The Geesaman fault dips steeply northeast in its western portion, and steeply to moderately southwest elsewhere. Features trending northeast are largely unaffected.

If the Stratton fault is a listric detachment, as suggested by its correlation with the San Pedro detachment, no further rotation of its lower plate is required for its restoration. Restoration of the fault brings the Stratton unit to horizontal.

Marble Peak-Lariat-Oro monoclines. The monocline-like nature of the restored Marble Peak syncline, coupled with the similar shape of the restored Lariat anticline and Oro syncline, suggests the possibility that all these structures are individual parts of Colorado-Plateau-style monoclines of Laramide age. The mutual relations of these folds to the Geesaman and Limb fault suggests that these monoclines grew along the faults. The Marble Peak-Lariat and Oro-Lariat monoclines may be the same feature, segmented by motion along the Geesaman fault. However, such correlation is difficult to square with deformation of the Geesaman fault by the Marble Peak syncline. Intrusion of Catalina Granite along the intersection of the monoclines (pl. 1) has made this question moot.

Restoration of a monocline rotates only the steep limb, rather than rotation of all limbs away from axial planes. Indeed, the impossibility of restoration of the folds in the normal manner is itself an indication that the folds form tilted monoclines; rotation of beds away from axial planes in this case produces either new folds or limitless tilted surfaces of unique orientation. In contrast, rotation only of the steep intermediate limb simply extends and joins supracrustal surfaces as restored for mid-Tertiary tilting.

Monoclinal restoration also involves the Geesaman and Limb faults and Leatherwood granodiorite. Movement on the Geesaman fault probably produced the oblique and discontinuous trends of elements of the Marble Peak-Lariat monocline, but further folding on the Marble Peak syncline portion has deformed the Geesaman fault (fig. 85). Restoration of the syncline thus produces a steeply south-dipping Geesaman fault and a splay thereof (fig. 92). Restoration of mono-

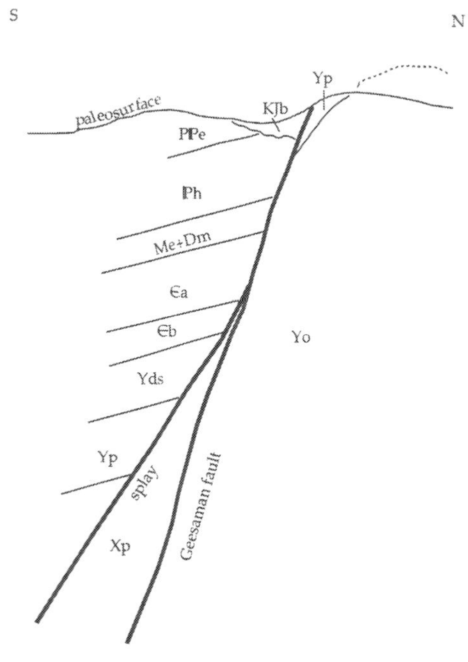

Figure 92 Cross-sectional diagram of restored relations at the east end of Geesaman fault. Units as for plate 1.

cline and fault together erases at least 1100 m of structural relief.

Similarly, restoration of the Lariat and Oro folds as a monocline implies restoration of the Limb and related faults. At least 200 m of structural relief are erased in restoring this system. Rice Peak porphyry configuration becomes as shown in figure 48. The proto-Mogul fault becomes a single plane.

Leatherwood granodiorite apparently intruded when monoclinal folding was about a third complete and when motion on the Geesaman fault was nearly complete. Restoration of the monocline thus implies that the Leatherwood intruded as a gently southwest-dipping sheet.

Proto-Mogul and Geesaman faults. These two faults as restored to this point are similar in the following respects: 1) WNW trend, steep dip, 2) left-oblique motion with north side up, 3) slip-vector indicators plunging gently east, 4) apparent basement offset of 7 km or more, 5) apparent dip-slip separation irregular but commonly 500 to 700 m (down to south), and 6) Laramide age. This similarity suggests that the two faults are related, as prominent members of a family of left-oblique faults on the north side of the Catalinas.

Restoration in such a way as to line up Oracle-Pinal basement boundaries (pl. 1B) would imply a total of about 14 km of left-lateral movement on the two faults. Such a restoration would also produce a straight Oracle-alaskite boundary (pl. 1B, and Bykerk-Kauffman, 1990, p. 49–50) and restore the Late Proterozoic diamictite, the dark lacustrine facies of the Dripping Spring Quartzite, the intermediate sandstone member of the Pioneer Formation, and perhaps the coarsest pyroclastic facies of Pioneer Formation, all to single belts or regions (pl. 1B). It is roughly consistent with permissible dip-slip separation magnitudes. A consequence of this restoration is the implication that the original northern part of the Catalinas now underlies parts of the Falcon Valley (fig. 93), but probably has been eroded down to basement by the up-to-north component of movement prior to burial. The rocks we do see in the northern Catalinas apparently moved there from the present area of the San Pedro basin (fig. 93). We can predict, for example, that old-variety Rice Peak porphyry, which predates the Geesaman fault, should be present south of the fault under cover of the San Pedro basin.

American Flag Formation, reverse faults, and related features. Restoration of features associated with the American Flag Formation can be done incrementally, as their relative ages are known. First we rotate the upper two members to horizontal through an average of about 20°. This figure is ambiguous due to a tendency for uppermost beds to dip more steeply. A gentle southwest dip on Proterozoic and supracrustal rocks in the immediate area results. However, the presence of clasts of Proterozoic rocks in the middle member, showing exposure of these rocks to the southwest, suggest this paleodip must reverse or be interrupted in that direction.

The faults bounding the American Flag Formation by this time dip about 50 to 60° to the southwest; it is for this reason that I call them reverse faults rather than thrusts. Restoration involves a modest elevation of the American Flag basin, but the uncertain stratigraphic position of tectonic inclusions along the fault plane make the magnitude of such uplift difficult to determine. Restoration of the American Flag structures also

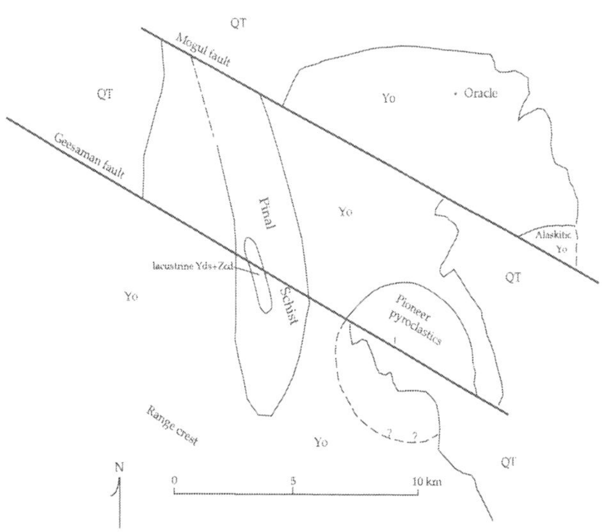

Figure 93 Map view of reconstruction along the proto-Mogul and Geesaman faults. Yo, Oracle Granite; QT, basin fill, shown for location but only locally offset.

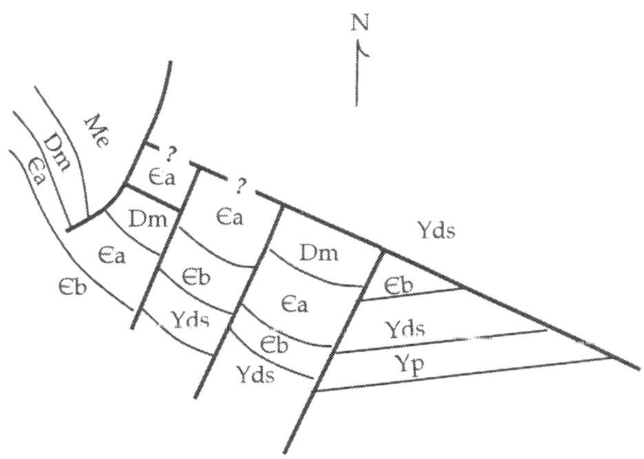

Figure 94 Pre-Rice Peak porphyry geologic map-diagram of the Stratton Wash area. Units from plate 1.

involves restoration of the gentle synclines of Rice Peak and Apache Peak, and the anticline of Peppersauce Wash.

The reverse faults described here and the Marble Peak-Lariat-Oro monoclines face in opposite directions tectonically; the area between is a high. Both sets of structures show evidence of having had a basement core. Therefore, the area between them must be considered a Laramide basement-cored uplift.

Features coeval with Rice Peak porphyries. The northeast-striking faults intruded by Rice Peak porphyry have been restored toward their southwestern ends by Janecke (1986), who showed that the faults restore themselves if the Rice Peak bodies are simply removed. Similarly, removal of the Rice Peak body toward the northeastern end of these structures restores stratigraphic continuity there, except as shown in figure 94. A few faults predating Rice Peak porphyry and perhaps a syncline remain. The apparent old fault in Stratton Wash may have been reactivated by the later Stratton detachment fault.

NORTH OF MOGUL FAULT

This area, the northernmost, consists mostly of Oracle Granite forming a high plateau on the south, and tilted Tertiary deposits to the north and east. Older supracrustal rocks have been stripped off by erosion. Laramide structural features are not apparent, though an attempt at Laramide reconstruction is made. The mid-Tertiary evolution of this area is the most intricate of the entire corridor. The Tertiary structures can be divided into four generations, summarized from recent descriptions by Dickinson, (1991, 1993), Force and Cox (1992), and Force and others (1995). Locality descriptions in Force and Dickinson (1994) are not repeated here.

Cloudburst-Turtle Detachment System

In the San Manuel-Mammoth area, the oldest structures are those of the Cloudburst-Turtle detachment system. The Cloudburst detachment fault is subhorizontal but irregular in detail, dipping about 4° to the east based on structure contours (fig. 95). It crops out as a somewhat sinuous trace (pl. 1) and in a total of four windows (only two in the study area) in the beds of deeply incised washes (fig. 96, locs. 94, 131). It separates Cloudburst Formation above, dipping steeply into the fault, from Oracle and related granitic rocks

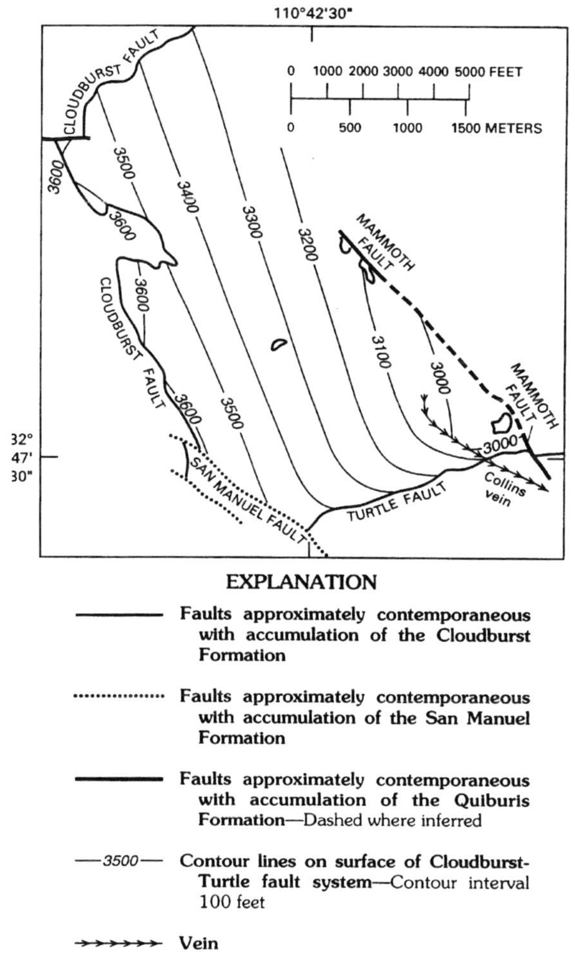

Figure 95 Structure contours on the Cloudburst detachment fault, from Force and Cox (1992).

Figure 96 Trace of Cloudburst detachment fault (Cf) in window of Tucson Wash (loc. 94). Tcbl, Cloudburst Formation; Yo, Oracle Granite.

below. The fault plane itself shows little gouge, but generally shows a chloritic breccia in the footwall. No "footwall ledge" is present, however, probably due to subsequent alteration. Cloudburst Formation on the hanging wall is an epidote-chlorite hornfels 50 to 90 m thick (because of tectonic juxtaposition with hot footwall; Force and Cox, 1992).

The Turtle fault (fig. 97, loc. 132) as originally mapped (cf. Creasey, 1965) separated Oracle Granite to the south from Cloudburst Formation to the north in the area of the San Manuel and Mammoth mines, i.e. east of the San Manuel fault. Dickinson (1987, 1991) and Force and Cox (1992) found that an offset segment of the Turtle fault extends west of the San Manuel fault and occupies a closely analogous position. In both segments, the dip of the fault plane is steep to the north (ranging from 50 to 70°). Epidotization and veining (by the Mammoth vein set) of the fault plane is common in the easternmost portions. In both segments, Cloudburst Formation is cut by the fault toward the west but overlaps it to the east. In the easternmost segment (loc. 57, 133), this relation is obscured by rhyolite and by alteration related to the Mammoth vein set (described by Force and Cox, 1992). Minor late motion on the Turtle fault cuts even the uppermost Cloudburst Formation.

The Cloudburst and Turtle faults are thought to be different aspects of the same structures (Dickinson, 1991; Force and Cox, 1992). Probably the Turtle fault served as a side-ramp of the Cloudburst detachment fault. Together they bound the Cloudburst or Tar Wash allochthon of Dickinson (1991), consisting of Cloudburst Formation that is cut at high angles by both faults but accumulated during fault movement. The northern boundary of the allochthon is beyond the map area (pl. 1).

Movement direction on the detachment system is revealed by the strike of its southern boundary, the Turtle side ramp, i.e. about N. 60 to 70° E. Top must have moved to the southwest, by a magnitude at least equal to the thickness of the Cloudburst Formation, about 3.5 km. Motion occurred concurrent with accumulation of Cloudburst Formation up to the horizons that overlap the fault. This interval gives ages of about 28 to 23 Ma, i.e. late Oligocene.

The detachment system cuts individual small bodies of rhyolite intrusive into Cloudburst Formation and Oracle Granite in the Tiger area (fig. 69), but does not displace the swarm at map scale, suggesting coeval faulting and intrusion. This suggestion is consistent with the presence of rhyolitic extrusives through the upper part of the lower member of the Cloudburst Formation, stratigraphically equivalent to beds that span the formation's overlap of the Turtle fault.

The extent of concealed parts of the Cloudburst detachment system is important for reconstruction. The presence of the Turtle fault bounding the Purcell window (pl. 1) on the north implies the presence of the Cloudburst detachment at depth north of the Turtle fault. To the west the Turtle fault is cut by younger faults, down to the west and/or south, and buried under younger deposits. If the concealed Turtle fault maintains its trend, it would remain buried to the western edge of the map area, possibly into the Falcon Valley.

Mammoth Vein-Fault

The Mammoth vein set (pl. 1) is developed along older faults striking NNW and dipping steeply west (loc. 57, 133). The main fault, called the Mammoth vein-fault, cuts the Turtle fault, but must postdate the Turtle only slightly (Force and Cox, 1992), as quartz of the vein is apparently derived from cooling rhyolite intrusives that are broadly coeval with the Cloudburst-Turtle detachment system. Force and Cox calculated dip-slip offset on this fault as 35 m (120 ft) down to the west, based on offset of the Oracle-Cloudburst unconformity. The distribution of vein width relative to fault strike (from Peterson, 1938) suggests some right-lateral

Figure 97 Turtle fault (Tf) in Tucson Wash looking east (loc. 132). Tcbl, Cloudburst Formation; Yo, Oracle Granite.

Figure 98 Photo of San Manuel mine (loc. 134) from south, showing San Manuel and other faults. Yo+Ks, mineralized bedrock; Tsmt, San Manuel Formation; SMf, San Manuel fault; Wf, West fault; Mf, Mammoth or East fault.

motion. The fault along the related east-dipping Dream and Collins East veins was also down to the west.

San Manuel and Coeval Faults

The San Manuel fault extends for over 12 km, across the study area from northwest to southeast (pl. 1). It dips about 25 to 30° to the southwest. Among the features it offsets are Laramide features such as the San Manuel porphyry system (Lowell, 1968), and older mid-Tertiary features such as the Cloudburst-Turtle detachment system and the Cloudburst Formation. Apparently it offsets the probable American Flag Formation also (loc. 55). The question of offset mineralized features is addressed by Force and Cox (1992) and Force and others (1995).

For much of the length of the San Manuel fault, the San Manuel Formation forms its upper plate (loc. 134, 135), and dips northeastward about 30° into the fault plane (fig. 98). In the north fork of Mammoth Wash, east-dipping San Manuel Formation is present on the lower plate of the fault also (loc. 136). The San Manuel Formation probably accumulated during fault movement, and large blocks locally accumulated against the fault plane (fig. 99). Toward the south end of the exposed fault, San Manuel Formation shows steeper dips adjacent to the fault, consistent with reverse drag.

Two splays of the San Manuel fault are shown on plate 1 for part of its length, modified from Force and Cox (1992). The southwestern splay of the fault apparently moved first, as the upper (Tucson Wash) member of the San Manuel Formation rests directly on older rocks to the northeast of that splay (loc. 136, 137). Dickinson (1991, 1993) showed this area slightly differently; neither interpretation is entirely satisfactory and the differences are virtually untestable with available exposures of closely similar formations.

The fault plane(s) of the San Manuel fault contains gouge as thin as 15 cm, commonly associated with coarsely crystalline black calcite (loc. 135). This black calcite forms clasts in adjacent San Manuel Formation. Locally the permeability of the fault has permitted copper from older deposits to migrate in solution and precipitate as oxides and/or carbonates. Slickensides and mullions on the fault trend S. 45° W. to S. 55° W.

Figure 99 Blocks of Cloudburst Formation and Tertiary rhyolite in basal Tucson Wash Member, San Manuel Formation (Tsmt), against the San Manuel fault (SMf), Tucson Wash (loc. 135). Yo, Oracle Granite.

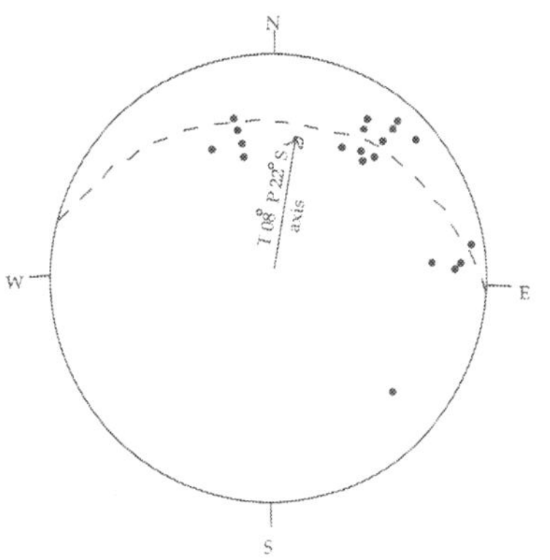

Figure 100 Stereographic projection of poles to bedding attitudes measured by Weibel (1981) from Cloudburst Formation between the Black Canyon and Red Rock faults, showing apparent fold axis.

Lowell (1968) gave the throw on the San Manuel fault as about 8000 ft (2440 m), with the western (top) side down and to the west along a vector trending S. 55° W. He based this figure on the offset of the San Manuel copper ore body. Force and Cox (1992) give corresponding figures of 2,360 m (7,750 ft) and S. 46° W., based on the offset of the line defined by the intersection of the Oracle Granite-Cloudburst Formation unconformity and the Turtle fault. Dickinson (1993) has subsequently shown that the position of this intersection is made somewhat ambiguous on the upper plate of the San Manuel fault by the Black Canyon fault (pl. 1).

Motion on the San Manuel fault was probably early Miocene in age, assuming the San Manuel fault and the San Manuel Formation to be coeval. The areal distribution of the San Manuel fault is not known, but the associated tilt-block domains are extensive, and include the unconformity of San Manuel Formation on Oracle Granite near the town of Oracle. Tilting of rocks on both the upper and lower plates suggests that the San Manuel fault originated as a steeper fault and was rotated to its present position (Dickinson, 1991).

The Black Canyon and Red Rock faults of Dickinson (1993) and Lowell (1968) respectively, are moderately to steeply south-dipping faults (Weibel, 1981) striking west to northwest. Movement is south-side-down, though a right-lateral component is known for the Black Canyon fault. Their kinematic relation to the San Manuel fault is not known. Both faults bound domains of regionally anomalous dip in Cloudburst and San Manuel Formations (pl. 1). Bedding attitudes measured by Weibel (1981) coupled with the shapes of the fault traces suggest that both these faults are folded (fig. 100). Dickinson (1993) noted that offset dies out structurally upward such that the contact between the Kannally and Tucson Wash Members of the San Manuel Formation is undeflected. This suggests that the two faults predate the Tucson Wash Member but postdate the Kannally Member, and are thus coeval with the San Manuel fault in a broad sense.

Basin-Range Faults

Numerous faults of the basin-range type generally separate the older units to the west from the Quiburis Formation filling the San Pedro basin to the east. They generally dip steeply northeast; their trends arc from northwest on the north to north-south on the south (pl. 1). These faults are adequately known only in the area of the Mammoth and San Manuel mines, where the two most important faults of this type are the Cholla and Mammoth faults. Toward the north, in the Mammoth district, these two faults merge (pl. 1). Force and Cox (1992) show that the total offset on the fault system in that area is 260 to 265 m (about 870 ft) straight down the steep easterly dip. Farther south, in the area of the San Manuel mine (loc. 134), the faults diverge. The Cholla fault to the east separates San Manuel Formation and San Manuel porphyry on the west from Quiburis Formation on the east. Its throw ranges from about 100 m on the north to 200 m on the south, down to the east, based on data in Schwartz (1953). The Mammoth fault, also known as the East fault in the mine area (fig. 98, loc. 134), separates Oracle Granite and San Manuel porphyry from San Manuel Formation at the land surface. Schwartz's subsurface data (contoured by Force and others, 1995) suggests offset of 100 to 200 m, again greater to the south.

The Mammoth fault cannot be traced south of the mine area, but the Cholla fault is intermittently exposed at least as far south as (new) highway 77, i.e. for a total of 4 km south of the bifurcation of the Cholla and Mammoth faults, and with the San Manuel fault outlines a long narrow bedrock horst extending to Smelter Wash, 7 km from this bifurcation. Other basin-range faults of the area include several known only in the San Manuel mine (Schwartz, 1953; Pelletier and Creasey, 1965; Thomas, 1966; Force and others, 1995), and faults

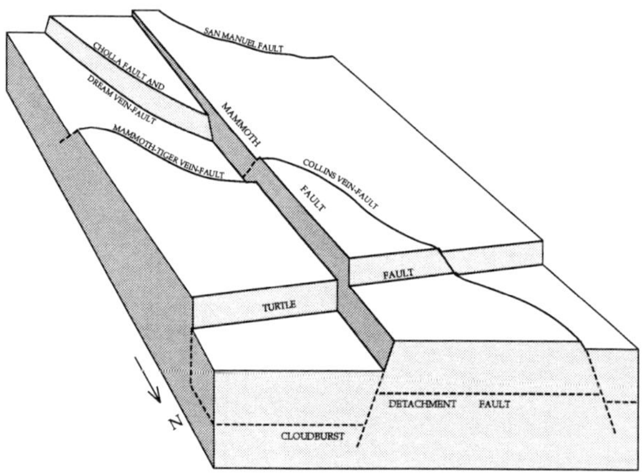

Figure 101 Schematic diagram of relation between structures in the Mammoth district (modified from Force and Cox, 1992).

separating bedrock from Quiburis Formation southeast of Smelter Wash (Dickinson, 1993).

Associated with the basin-range faults is the Quiburis Formation of post-mid-Miocene age; it probably accumulated as faulting occurred. The Quiburis Formation now dips eastward 5 to 10°. Lack of dates on Quiburis Formation prevents conclusions on the local age of basin-range faulting.

Structural Restoration

Incremental reconstruction from younger to older of the rather well-known episodes of faulting and associated deposition results in the following sequence. Visualization of reconstruction is aided by figure 101.

Removal of basin-range faulting and Quiburis Formation may have no tilting consequences in the bedrock part of the area, as the dip of Quiburis Formation may be a basin-subsidence feature. Restoration of

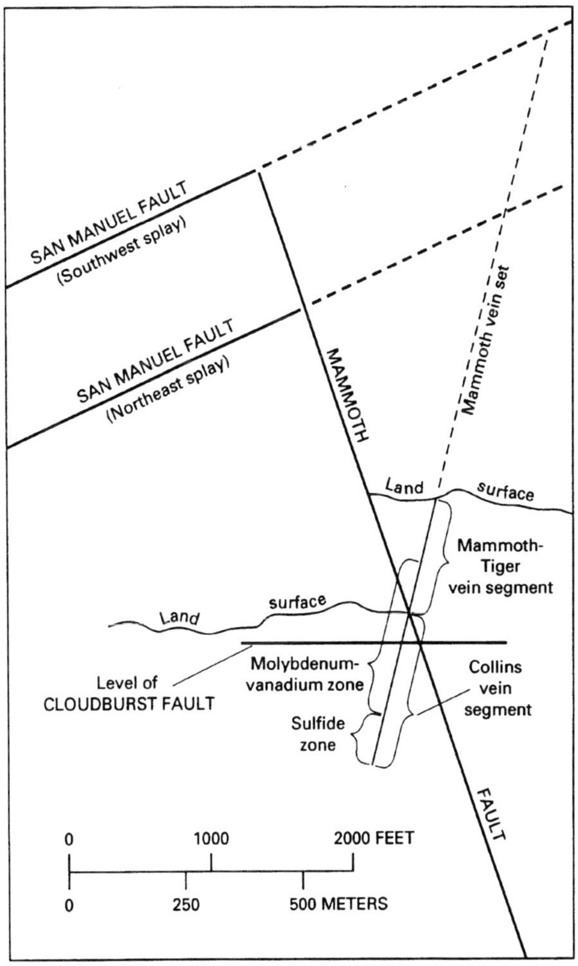

Figure 102 Cross-sectional diagram of reconstructed Mammoth vein set showing mineral zones (from Force and Cox, 1992).

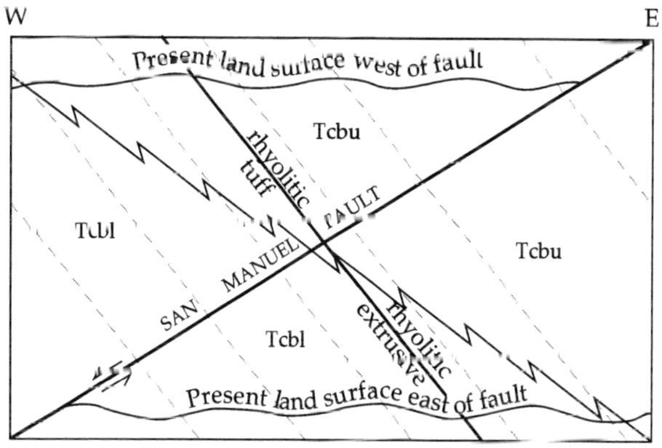

Figure 103 Cross-sectional diagram of reconstructed Cloudburst Formation, modified from Force and Cox (1992, fig. 18).

the Mammoth fault joins the segments of the Mammoth vein and vein-fault to a single plane, with mineral assemblages zoned as in figure 102.

Restoration of the San Manuel fault and removal of San Manuel Formation requires tilting down to the southwest through a mean angle of 30 to 35° about an axis trending N. 40° W. (Force and others, 1995), apparently over the entire area being considered, including the Oracle area. The effects of this stage of restoration are a) to increase the dip on the Cloudburst detachment to about 34°, b) to rotate the Turtle fault (east segment) to a strike of N. 50° E., dipping about 54° N., c) to bring the Cloudburst Formation to mean northeastward dips of 10 to 20° in the area north of the Turtle fault, and horizontal south of it, with a sub-horizontal unconformity on Oracle Granite, d) to rotate the "tablocylindroidal" San Manuel ore shell and associated productive portion of San Manuel porphyry to a strike of N. 81° E. and a southeast dip of 62° (Force and Dickinson, 1994; Force and others, 1995), e) to rejoin the segments of the porphyry copper ore body and related porphyry dikes (Lowell, 1968), the Oracle-Cloudburst unconformity, and the Turtle fault, with the restored attitudes listed above, for a NE-SW horizontal shortening of the study area by about 2050 m (6700 ft), f) to rotate the Proterozoic diabase dikes of the region to a northeast dip of about 10 to 20°, and g) to rotate the Mammoth vein-fault plane to the attitude of a high-angle reverse fault dipping steeply northeast.

Restoration of the San Manuel fault also constrains some original features of Cloudburst Formation. The magnitude of separation suggests correlation of the rhyolitic tuffs of the southwestern area with the rhyolitic volcanics (including welded ash-flow tuffs) of the northeastern area, as might be expected on volcanologic grounds. This in turn requires marked time-transgressive facies change between the upper and lower members of the formation (fig. 103). The sense of implied interfingering is such that the lower member probably has a component of supply from the east. Provenance considerations suggest this component is important only in the northeastern area.

Restoration of the Mammoth vein-fault itself elevates the eastern side 35 m and moves it an unknown distance to the north, relative to the rest of the area in the Tiger region. Offset on the Dream-Collins East structure possibly doubles this figure (Force and Cox, 1992).

Restoration of the Cloudburst-Turtle detachment system and removal of Cloudburst Formation strongly affects only the area of the upper plate (Cloudburst allochthon), i.e. the area north of the Turtle fault (Force and others, 1995). The lower plate, the area south of the Turtle fault, was apparently not tilted and needs no restoration (except removal of a thin cover of Cloudburst Formation). Restoration of the area north of the Turtle fault requires moving it at least 3.6 km northeast, and tilting it through its remaining dip up to the northeast.

The rotated allochthonous domain trends southwestward toward the Mogul fault. The Cloudburst and Turtle faults may thus be present in the area south of the neo-Mogul fault where late Tertiary left-lateral strike-slip faulting could have brought them back toward the study area. However, correlation with the Stratton detachment seems unlikely, as about 15 km of offset are required.

At this point in reconstruction we are dealing with a Laramide configuration of features, the one pertinent to intrusion of San Manuel porphyry. For reconstruction of Laramide features themselves we can work with only the probable American Flag Formation in Mammoth Wash and the few pre-Laramide features of this and nearby areas. The 34° mean dip of probable American Flag Formation is sufficiently gentle that little additional rotation is required. Jon Hagstrum (in Force and others, 1995) finds that the San Manuel porphyry itself has been tilted about 33°, based on its paleomagnetism.

Force and others (1995) show that originally horizontal pre-Laramide features of this tilt domain now show mean northeast dips of about 45°. Within the study area, the only such features are diabase intrusives in Oracle Granite, which originally were sub-horizontal sills (Howard, 1991). The uncertainty of the exact original dip of these bodies is such that for our Laramide reconstruction we can only surmise a gentle northeast dip.

Laramide reconstruction of the area also involves the proto-Mogul fault, which seems to have had over 7 km of left-lateral motion. Restoration still leaves the Oracle block high, as the probable American Flag Formation, which elsewhere predates the proto-Mogul fault, rests directly on Oracle Granite. The original configuration of the American Flag Formation across the fault is unclear, due to near-parallelism of sedimentary dip and the fault vector.

Figure 104 Quartz-adularia-hematite veinlets related to Mammoth vein set in Tertiary rhyolite host, Tucson Wash (loc. 132).

Figure 105 Sub-horizontal faulting along a horizon that may be a paleosol on Dripping Spring Quartzite. Opening to workings of Southern Belle mine at lower left.

Table 10 Mineral properties and prospects of the study area. Undescribed and barren workings omitted.

District name Deposit or prospect	Production years	Production (ore) tons	Grade	Form	Hosts (pl. 1)	Age	Ore minerals	Gangue minerals	Control	Alteration	References*	Locality numbers (others on base map)
Mammoth	1879-1959, 1978-	2×10^6	1-15 ppm Au, Mo, V., Pb, Zn	quartz veins	Yo, Tel	about 23 Ma	Au, wulfenite, vanadinite, galena, sphalerite	adularia, hematite, amethyst, Mn ox, barite, fluorite	Mammoth vein faults, Tertiary rhyolite, Cloudburst detachment	epidote chlorite	Peterson, 1938; Force and Cox, 1992	57, 133
Tar(Tarr)	?-1943	minor	?	quartz veins	Yg	mid-Tertiary	scheelite, wolframite, Cu mins		Cloudburst detachment	chlorite	Wilson, 1941; Creasey, 1965	
San Manuel (+Kalamazoo)	1955-	3×10^6/yr	0.8% Cu	disseminated	Yo, Ks	about 64 Ma	chalcopyrite, molybdenite, bornite		tabular intrusion, ?Cretaceous unconformity	zoned K-feldspar to propylitic	Schwartz, 1953; Pelletier and Creasey, 1965; Lowell, 1968, Force et al., 1995	72
American Flag												
American Flag	1878-1961		Cu, Ag, Pb, Au	silicified veins	Ydb, Yo	?	chalcopyrite, galena, Au	Fe-carbonate	faulted contact			139
Bearcat	1939-1943		W, V, Mo, Au, Ag	quartz veins	Ydb, Yo	?	scheelite, vanadinite, wulfenite	calcite, Fe ox, pyrite	faulted contact			140
Hijinks			Au	quartz veins	Ydb, Yo	?	Au, scheelite		faulted contact			27
Magrugador	1940-1980		Au	quartz veins	Yo	?	Au, galena		fault			141
Campo Bonito												
Morning Star	1913-1953	>6000	up to 10% WO3	fracture	Me	?	scheelite, $PbCO_3$		Mogul fault		Bateman and Erickson, 1944; Dale, 1959	
Pure Gold	1943-1953	about 6000	up to 10% WO3	fracture	Me	?	scheelite		Mogul fault		Bateman and Erickson, 1944; Dale, 1959	
Southern Belle												
Southern Belle	1885-1964	>18,000	Au, Ag, Pb	quartz veins	Yds, Zc	?	Au, galena, cerussite, scheelite	pyrite	sheared unconformity			
Southern Belle placers	1884-		Au			modern	Au		Southern Belle lodes, Mogul fault		Johnson, 1972	
Old Maudina	1915-1916 & 1943-1944		W, Mo, Pb	vein	Ca	?	scheelite, wulfenite, cerussite		fault		Wilson, 1941; Creasey, 1967	
Mary West	1955-1960		Pb, Cu, Zn, Au, Ag	quartz veins	Yp	Laramide	galena	pyrite, siderite, epidote	fault, sedimentary contact			118
Geesaman fault												
Taylor X (Catalina Camp)	?-1953	minor	W, Cu	quartz veins	Ca	Tertiary?	scheelite, chalcopyrite		Geesaman fault	epidote, chlorite		
Corregedor	1940's	minor	4% WO3		Yp pyroclastic	Tertiary?			Geesaman fault			14
Marble Peak												
Geesaman & Daily	1882-	115,000	2-3% Cu	skarn	Ca, Dm, Me	Laramide	chalcopyrite, bornite, chalcocite, galena, sphalerite, scheelite	garnet, epidote	Leatherwood granodiorite, old faults		Peterson and Creasey, 1943	148, 149
Leatherwood	1939-1968	3000	5-12% Cu	skarn	Ca, Dm, Me	Laramide	chalcopyrite, bornite, galena, sphalerite	garnet, epidote, magnetite	Leatherwood granodiorite, old faults		Peterson and Creasey, 1943	
Stratton		minor		skarn	Ca, Dm	Laramide	chalcopyrite, molybdenite		Leatherwood granodiorite		Peterson and Creasey, 1943	
Hartman-Homestake	?-1948	600	3% Cu, 3% Zn, 5% Pb	skarn	Ca, Dm, Me	Laramide	chalcopyrite, sphalerite, galena	garnet, epidote	Leatherwood granodiorite, old faults		Peterson and Creasey, 1943	

* Additional publicly available but unpublished sources are 1) the Mineral Resource Data System (MRDS) of the U.S. Geological Survey, incorporating data from many sources, 2) a 1945 report by E. J. Ewing on the Campo Bonito district, available from the Buffalo Bill Historical Center, Cody, Wyoming, and 3) files of the Anaconda Collection, available from the American Heritage Center, Laramie, Wyoming.

MINERAL RESOURCES

Mineral districts of the study area include (in approximate order of importance) San Manuel (+Kalamazoo), a major porphyry copper district of Laramide age; Mammoth, formerly a major district of mid-Tertiary vein deposits containing gold, molybdenum, and other metals; Marble Peak, a Laramide copper skarn district currently in production; Campo Bonito, a tungsten-gold district; and Southern Belle, a gold deposit. All these deposits are north of the range crest; indeed all but Marble Peak are north of the Geesaman fault. They are described approximately from north to south; a full description forms appendix 1, and a summary forms table 10. Mines not labeled on the topographic base are shown as numbered localities (pl. 1) and keyed to appendix 2.

DISTRICT SUMMARIES

Mammoth District

Mid-Tertiary veins of the Mammoth set were probably the first exploited in the study area. They yielded gold, silver, molybdenum, vanadium, lead, zinc, and more recently silica from about 2 million tons of ore from 1879 intermittently to the present. The deposits are closely associated with mid-Tertiary rhyolite, and hosted by Oracle Granite, the unconformably overlying lower member of Cloudburst Formation, and rhyolite (pl. 1). The mineralized veins are rather distinctive (fig. 104).

The veins, dipping steeply southwest, have been offset by the northeast-dipping Mammoth fault zone into two segments called the Collins (loc. 57) and Mammoth-Tiger (loc. 102) segments (Peterson, 1938; Creasey, 1950; Force and Cox, 1992). Offset has apparently been about 265 m down the fault dip. Reconstruction of the vein (fig. 102) produces a zonation over 550 m of vertical relief from lead and zinc sulfides at lower levels, through a wulfenite-vanadinite zone, to gold and possibly supergene assemblages at upper levels. Mineral zonation seems to center near the intersection of the vein set with the Cloudburst detachment fault. The form of relations among mineralization, detachment, and rhyolite intrusion is explored by Force and Cox (1992), in appendix 1, and the section by Kamilli (this volume). Potential for further mineralization in this district seems greatest along the Cloudburst-Turtle detachment system.

San Manuel (+Kalamazoo) District

The San Manuel porphyry copper deposit and its offset segment the Kalamazoo deposit are Laramide (apparently about Ma) in age. The former was discovered in 1943–1944 and the latter in 1965–1966. Except for the Red Hill discovery outcrop, the entire system is concealed. The system has been a major producer since 1955 at an average copper grade of about 0.8 percent.

The host rocks of the deposit are the Oracle Granite and San Manuel porphyry (fig. 98). The orebody forms a shell in and around one intrusion, and associated alteration zones are concentric (Lowell, 1968; Lowell and Guilbert, 1970). Mid-Tertiary offset on the San Manuel fault segmented the two parts of the original orebody. A discussion of the geometry and tilting history of the deposits is in appendix 1 and is amplified by Force and others (1995), Guilbert and Lowell (1996), and Dickinson and others (1996). Potential for further mineralization of this type nearby may be greatest where the Oracle-American Flag unconformity is intruded.

American Flag District

Deposits hosted by Oracle Granite between the Mogul fault and Oracle have been grouped as the American Flag district. Most deposits are along sheared contacts between Oracle Granite and Proterozoic diabase. Most were worked for gold but some for W-Mo-V or Cu-Ag-Pb; all the deposits are small. Their ages are problematic; detailed description is in appendix 1.

Campo Bonito District

Deposits directly tied to the Mogul fault in the study area have been grouped as the Campo Bonito district. The most important deposits contain scheelite and are in a single fault-bounded block of Escabrosa Limestone along the Mogul fault. Further potential seems limited.

Southern Belle District

Deposits south of the Mogul fault but probably related to it have been grouped as the Southern Belle district. Host rocks are sedimentary, ranging from Apache Group to Paleozoic. They contain scheelite in carbonate host rocks and gold-silver-lead in non-carbonate hosts. Mineralization is in shear zones parallel to bedding (fig. 105). Some potential for further deposits may be related to the Bolsa-Dripping Spring unconformity and the oxidizing capacity of the locally intervening Campo Bonito unit.

Deposits along the Geesaman Fault

A zoned population of deposits and occurrences along the Geesaman fault and a deformed splay thereof are treated herein like a district. They pass from greisens on the west through tungsten deposits into base- and precious-metal deposits on the east. They are thought to be related in age and origin to Catalina granite. Apparently they have never been explored as a single population.

Marble Peak District

Deposits of this district were worked as early as 1882 and have been intermittently active to the present. The deposits are skarns related to intrusive contacts of the underlying Leatherwood granodiorite intrusive sheet with Paleozoic carbonate host rocks (fig. 106). Lithologically, the skarns are garnet-pyroxene rocks with retrograde epidote and sulfide minerals. The main ore minerals are chalcopyrite and bornite. Lead, zinc, gold, silver, and other metals have been recovered locally. The stratigraphy of host rocks and pre-Leatherwood faults (pl. 2B) control details of the size and character of mineralization. The skarn bodies broadly follow the Leatherwood contact but in detail form "Christmas trees" rooted in the contact (fig. 72), especially at intersections of older faults. Subsurface extensions of these faults may have potential for additional deposits. Potential for additional skarn deposits, especially in Abrigo Formation, is widespread on the north-facing slopes of the main range (pl. 1).

Non-Metallic Deposits

Marble and slate localities near the Geesaman fault may have economic potential. Other possible non-metallic commodities in the study area are also described in appendix 1.

AGE AND ORIGIN OF DEPOSITS

Mineral deposits of the study area are of probably pre-Laramide, Laramide, and mid-Tertiary age. The probable pre-Laramide deposits are in the American Flag district.

Laramide deposits include the San Manuel porphyry copper deposit and the Marble Peak skarn deposits; both are clearly related to Laramide plutons. Other skarn deposits related to Leatherwood granodiorite are also Laramide. The Mary West deposit (appendix 1), which apparently is related to Rice Peak porphyry, is probably Laramide. The Nugget Canyon prospect (appendix 1) is possibly Laramide.

Mid-Tertiary deposits in the study area include the Mammoth district, and prospects such as the Shultz Spring altered area (appendix 1) that are hosted by Cloudburst Formation. Prospects that postdate mylonitization, such as those of the Catalina forerange and the Valerie May prospect (appendix 1), and those that are along the San Manuel fault, must be mid-Tertiary or younger.

The ages of the Campo Bonito and Southern Belle districts along and south of the proto-Mogul fault are not clear, as they are not clearly related to any intrusive igneous rocks, or other clearly dated feature. The deposits of the American Flag district, though probably pre-Laramide in part, probably includes deposits related to the Mogul fault also.

The ages of deposits along the Geesaman fault are also unknown, but for them a strong case for mid-Tertiary age can be made. The Reef of Rock facies of Catalina granite (of mid-Tertiary age) intrudes the western part of the fault plane as aplite dikes, and related greisens are found in the fault plane at least as far east as Dan Saddle. Moving progressively farther east, tungsten deposits and then base- and precious-metal deposits are found. The zonal arrangement suggests a relation to the mid-Tertiary Catalina granite toward the west

If the Geesaman and proto-Mogul faults had similar motions and are of about the same age, as

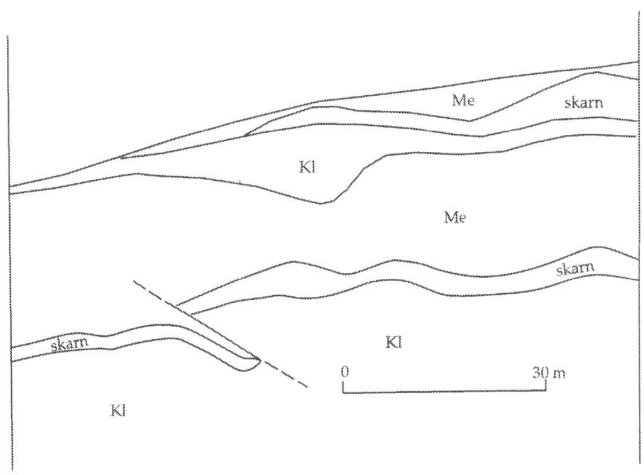

Figure 106 Cross-sections of asymmetric skarns from Daily mine (loc. 150), from Peterson and Creasey (1943). Kl, Leatherwood granodiorite; Me, Escabrosa Limestone.

argued in previous sections, then it seems likely that some mineralization along them is the same age. That is, if the deposits along the Geesaman fault are mid-Tertiary, some along the related Mogul fault are probably mid-Tertiary also. All the known relations permit this interpretation. The Catalina granite may have intruded the proto-Mogul fault west of the map area. Mineralization along the Mogul fault is concentrated toward the west except in areas of extremely receptive host rocks and close-spaced cross-fractures (Campo Bonito and part of Southern Belle districts). Perhaps all the deposits with wulfenite-vanadinite and(or) scheelite ores (including Bearcat-type deposits of the American Flag district) are mid-Tertiary and the gold-dominant ores are older.

The more important districts of the study area owe their origins either to the effects of intrusion, to the plumbing systems along extensive faults, or to combinations of the two. The Laramide deposits seem to be more related to intrusions; some are porphyry copper deposits (and probably-related occurrences) and some are skarn deposits. The mid-Tertiary deposits seem to be related more to extensive faults, some of which are mid-Tertiary detachments and some of which are older left-oblique faults. Relations to felsic intrusives can also be discerned for the mid-Tertiary deposits; this is most obvious at Mammoth, where Tertiary rhyolite is associated with mineralization, and along the Geesaman fault, where hydrothermal zonation apparently proceeds outward from an intrusive contact with the leucocratic Reef of Rock facies of Catalina granite.

DISTRIBUTION OF MINERALIZATION

Areal

Three conclusions are apparent from an areal view of known mineralization in the study area. First, two faults that both were originally left-oblique with large offsets, the Mogul and Geesaman faults, served as the plumbing systems for widespread mineralization. In the case of the Mogul fault, related deposits probably include the Campo Bonito, Southern Belle, and part of the American Flag districts inside the study area, and the Little Hill and probably the Burney districts outside the study area, with associated placer deposits. Along the Geesaman fault, the Taylor X, Corregedor, and Bluff deposits (appendix 1), associated placer deposits, and several unnamed deposits, are fault-related. In addition, veins of greisen follow the Geesaman fault toward its western end. Mineralization along both the Mogul and Geesaman faults is dominated by gold and tungsten, with some deposition of lead, copper, molybdenum, vanadium, and silver. Along the Geesaman fault, there is a well-defined zonation from greisen on the west through tungsten deposits to base- and precious-metal deposits on the east.

Second, the basement-cored uplift bounded by the Marble Peak-Lariat-Oro monocline and the American Flag reverse faults is relatively devoid of Laramide mineral deposits. The laccolithic older type of Rice Peak porphyry is coincident with this uplift and generally lacks the thermal and alteration effects of areas intruded by either the Leatherwood granodiorite or San Manuel porphyry.

Third, within each of the three northeast-tilted crustal blocks separated by the Geesaman and Mogul faults, Laramide mineralization is concentrated along the northeast edge, where crustal depths at the present land surface were less at the time of mineralization, and where host rocks formed effective traps. In contrast, probably mid-Tertiary mineralization is concentrated along the northeast edge of the entire study area, except for deposits along the Geesaman, Mogul, and Catalina faults. Laramide paleodepths vary within individual blocks due to later block rotation. In the mid-Tertiary the range started behaving as a single flexing block to some extent, so that shallow paleodepths were available only at the northeast end, and later at both extremities.

Thus, in the block south of the Geesaman fault, Laramide mineralization along the northeast edge comprises the Marble Peak district. To the southwest, isoclinal folds were forming at the same time and few deposits are known. Mid-Tertiary mineralization is common only along the Geesaman fault itself. In the block between the Geesaman and Mogul faults, Laramide mineralization along the northwest edge comprises the Mary West deposit. To the southwest, basement rocks are exposed and few mineral deposits formed at this time. The Campo Bonito and Southern Belle districts and part of the American Flag district may be mid-Tertiary, formed along the Mogul fault. In the northern block, north of the Mogul fault, both the Laramide San Manuel and mid-Tertiary Mammoth districts are far to the northeast. For the San Manuel district, probable American Flag Formation implies a shallow Laramide environment. Similarly, for the Mammoth district, the Cloudburst Formation implies a shallow mid-Tertiary environment.

Stratigraphic

In stratified hosts, a stratigraphic control of mineralization is apparent; it commonly follows the lowest chemically reactive horizon (fig. 107). Depending on solution composition, precipitation may be by Eh, pH, or other chemical change, and the reactive horizons vary accordingly. In the study area, the lowest reactive horizon for a given reaction type also varies from place to place as a result of lateral change. For example, the lowest precipitant for an acidic solution would be Mescal Limestone in a few sections, but the Abrigo Formation in most. Probably, the effect of a reactive formation is increased if an adjacent formation is an aquifer.

SUMMARY OF MINERAL RESOURCE POTENTIAL

Potential for additional mineral resources, described in appendix 1, may be associated with the following features (from north to south): 1) the hanging wall of the Cholla fault in the San Manuel area, 2) the sub-American Flag unconformity as a control of porphyry deposits, 3) the intersection of the Cloudburst detachment with the Turtle fault, 4) alteration in Nugget Canyon, 5) the sub-Campo Bonito unconformity, 6) mineral zoning along the Geesaman fault, 7) pre-Leatherwood faults in the Marble Peak district, 8) voluminous Abrigo Formation, commonly contact-metamorphosed, north of the range crest, and 9) the Catalina detachment fault.

SYNOPSIS

STRATIGRAPHY

The study area contains a basement of Pinal Schist and Oracle Granite, overlain by a section of Apache Group (Middle Proterozoic) through Paleozoic rocks as thick as 1250 m. This sequence is much like those of neighboring ranges and of previous descriptions except in the following respects:

Pinal Schist is far more abundant in the study area than previously thought; it forms about 20% of the basement, as a N-S elongated belt, offset by the Geesaman and Mogul faults. In the study area, fine clastic rocks and rhyolite are subequal in abundance as protoliths of Pinal Schist.

Oracle Granite shows an alaskitic phase north of the Mogul fault in pediments adjacent to the San Pedro basin. Phanerozoic deformation fabrics increase southward such that in the forerange, thick screens of Oracle between Eocene leucogranite sills are mylonitic. Chemical, textural, age, and other evidence confirms that this material is indeed Oracle Granite.

In the Apache Group, Pioneer Formation includes a coarse pyroclastic facies, including silicic ash-flow tuff, centered just north of the Geesaman fault. These pyroclastics are mostly in the upper part of the formation, but where they are as thick as 100 m they may rest directly on Scanlan Conglomerate Member.

The Dripping Spring Quartzite is probably nonmarine, based on the lenticular nature of the Barnes Conglomerate Member, and sedimentary structures in overlying quartzite and argillite. The dark argillite is probably lacustrine. Mescal Limestone is present in only a few parts of the area, and is represented by about 50 m of lithotypes that are basal in Mescal sections elsewhere.

The Campo Bonito Formation of probable Late Proterozoic age partially fills steep-walled paleovalleys cut through Mescal Limestone into Dripping Spring Quartzite. Its thickness is about 20 m, of hematitic siltstone and/or diamictite. The diamictite is probably glaciomarine, and the hematitic siltstone is probably related to Rapitan-type iron-formation.

Bolsa Quartzite varies in thickness from about 10 to 100 m, locally over short distances. Thinner sections tend to rest on older parts of the Apache Group, indicating a draping effect over pre-Bolsa structure-controlled topography.

Abrigo Formation is more abundant than previously thought in the study area; it forms large areas just north of the range crest. The Abrigo was leached of calcium in the area between the Geesaman and Mogul faults. Near the range crest, calcium was baked in as calc-silicates by Laramide metamorphism.

The Mt. Lemmon unit, probably Lower or Middle Devonian in age, is fluvial, deposited in a sinuous

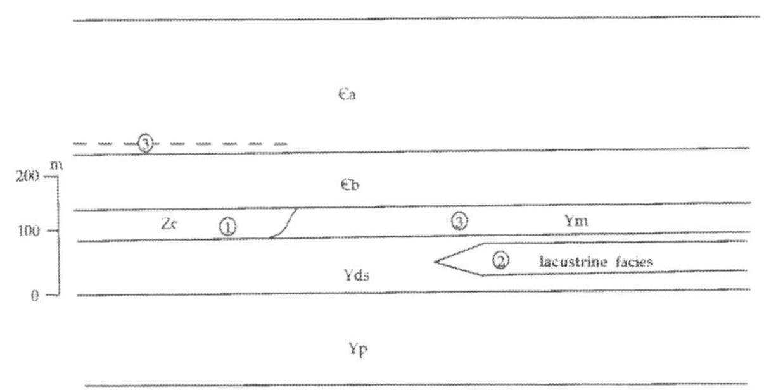

Figure 107 Stratigraphic diagram of inferred stratigraphic-geochemical loci of mineralization in the study area, assuming ascending solutions that precipitate at the lowest reactive horizon. 1, lowest oxidizing horizon; 2, lowest reducing horizon; 3, lowest alkaline horizon. Unit designations as for plate 1.

paleochannel cut through upper Abrigo Formation in the Mt. Lemmon area. Relations to Martin Formation are not exposed, but the Mt. Lemmon unit probably correlates with the unconformity at the base of the Martin.

The Naco Group in the study area is represented only by the lower portion, of 260 m thickness. The Earp Formation, the highest formation present, contains a thick quartzite unit.

Probable Bisbee Group is present along the Geesaman fault, but it is more restricted than previously thought. It may or may not be related to the Geesaman fault.

American Flag Formation can be divided into three members that record the progression of Laramide events. Uplift of a NW-trending part of the range along reverse faults caused erosional stripping first of Paleozoic, finally of Precambrian basement rocks. In the same time span, both types of Rice Peak porphyry were intruded, then the older type was exposed to erosion. Probable American Flag Formation extends northward almost to Mammoth, and rests on Oracle Granite. The depositional geometry relative to synchronous faulting suggests a paleo-San Pedro foreland (?) basin of Laramide age.

Mid-Tertiary units on the northern margin of the study area aggregate about 5500 m in composite thickness of coarse clastic and volcanic rocks. Those on the southern margin are as much as 3500 m thick. They represent the fill of basins left by detachment faulting and low-angle normal faulting, and far exceed the thickness of pre-Tertiary supracrustal rocks of the Santa Catalina range itself. These units (Cloudburst and San Manuel Formations on the north, Pantano Formation and Tinaja beds on the south) have been described elsewhere, but the Stratton unit on the San Pedro margin within the study area is new. It also consists of coarse clastic and volcanic rocks, and rests on the Stratton detachment fault. Its exposed thickness is small, but total thickness is probably over 600 m.

INTRUSIONS

Proterozoic diabase forms sills aggregating about 300 m in thickness in Apache Group, and formerly subhorizontal sheets also aggregating about 300 m in Oracle Granite.

Late Cretaceous Leatherwood granodiorite forms a 300 m subhorizontal sheet, emplaced as shallow as 1.9 km at its northern end. The epidote formerly thought to be magmatic is here interpreted as secondary. The Leatherwood sheet shows an abrupt lateral termination in the valley of Alder Creek. The sheet was emplaced by wallrock dilation, and the termination appears to correspond to a syn-intrusion fault cutting only sequences above the sheet. The sheet cuts structurally across section from the level of Escabrosa Limestone against the Geesaman fault to the north, to Oracle Granite at the range crest to the south, engulfing several faults along the way. Still farther south, Leatherwood forms two intrusions in Oracle Granite in the forerange, where both rocks are mylonitic. Near the range crest, Leatherwood shows a strong Laramide fabric.

The intrusive relation of Leatherwood to several major faults makes it possible to relate Laramide land-surface phenomena (American Flag sediments, vent-facies Rice Peak porphyry) to Laramide paleodepths in the area of the range crest.

Rice Peak porphyries include two rocks of different ages. The older is pre-Leatherwood and pre-upper American Flag, whereas the younger is post-Leatherwood and post-lower American Flag. The older type is the more voluminous, and occupies an area north of Leatherwood granodiorite intrusion. The porphyries form a sill or laccolith at the base of the Apache Group, with less-concordant appendages above and below. Aggregate thicknesses average about 250 m. Rice Peak porphyries are associated with remarkably little contact metamorphism or hydrothermal alteration, apparently because of many small intrusive pulses. Rice Peak intrusion inflated the supracrustal section, especially in two areas where intrusion produced an intricate puzzle of screens and xenoliths.

San Manuel porphyries include an older "type B" broadly concordant to regional dip, forming a "Christmas-tree" sill, and intruded in part along the Oracle-American Flag unconformity. The younger "type A", the body related to porphyry copper ore, formed a dike 300 m thick.

Seven mappable leucogranite intrusions of Eocene age (Wilderness suite) form an aggregate thickness of 3.9 km south of the range crest, and were intruded into an interval of crustal section originally only 0.7 km thick, centered on the Oracle-Pioneer unconformity. Two of the sills, with a total thickness of 2.8 km are garnet-bearing and muscovitic, but all are peraluminous. The thickest sill at about 2000 m, the Wilderness granite, is emplaced along the Oracle-Pioneer unconformity. Above it is the Spencer composite sill, about 400 m thick; below it in descending order are the Catnip Canyon banded granite (300 m), Shreve Pass (60 m), Thimble Peak (800 m), Gibbon Mountain (240 m), and Seven Falls (>150 m) sills. The Wilderness granite crosses into the forerange and provides the vehicle for tectonic reconstruction. At its top is a locally developed pegmatite-aplite zone several hundred meters thick, composed toward the base of a horizontal sheet but composed toward the top of steep dikes. Some dikes of intermediate age are dominated by aplites with garnet bands. Mineral assemblages of the pegmatites are simple.

Mid-Tertiary intrusives in the study area include the Knagge granite and the Catalina granite and associated ring-dikes in the range-crest area. These may be voluminous at depth under the range. The Catalina granite intrudes the Geesaman fault and has greisenized one segment of it. Small rhyolitic and rhyodacitic intrusive bodies occur in the northern part of the study area. The rhyolite also forms extrusives in Cloudburst Formation.

In sum, largely-concordant plutons of various ages have inflated the 1250 m supracrustal section and its basement by about 900 m throughout most of the study area, and by as much as 4800 m in the southern part of the area.

METAMORPHISM AND FABRICS

Pinal Schist shows low metamorphic grades and a steep north-trending phyllitic fabric as a result of Precambrian metamorphism. Segregation of quartz to form schist occurred only locally.

Laramide fabric is strong in the range crest area, and is present though overprinted south of the crest. The fabrics consist of phyllitic to schistose foliation and several types of lineations, commonly refolded. Much fabric postdates Leatherwood granodiorite and older Rice Peak porphyry, and some affects Spencer granite, but in some areas fabric elements are cut by Leatherwood. Thus fabrics apparently range in age from Late Cretaceous to early Eocene. Metamorphic grades are low except where the Leatherwood has caused skarn or other high T/P metamorphism. However, paleodepths of only about 5 km in this area suggest high Laramide geothermal gradients.

Intrusion of Wilderness suite leucogranites caused only local low-grade metamorphism and except for Spencer granite accompanied no fabric development; commonly greenschist-facies assemblages occur with crenulated Laramide fabrics in pelitic rocks adjacent to Wilderness intrusive contacts. Most host rocks, however, show neither fabrics nor metamorphism adjacent to Wilderness intrusions. Much metamorphism previously thought to be Wilderness-related occurs where Leatherwood intrusion preceded Wilderness intrusion.

Mid-Tertiary mylonitization is pervasive in all the rocks of the forerange, and some on the south face of the main range. Pure shear was a component of mylonitic strain. Two stages of mylonitization are apparent; an early (pre-26 Ma) top-to-southwest stage formed a 600 m thick belt, floored by fully-ductile rocks, that was folded to form the Forerange anticline. Growth of the anticline must have made part of this belt inefficient for further movement; the later stage included 1) a continuation of the same motion along the south limb of the anticline, 2) a continuation of this motion along the structurally higher Windy Point zone (post-26 Ma) above the northern limb of the anticline, and 3) reversed motion (top-to-northeast) in the original belt on the northern limb of the anticline (fig. 78). The geometry gives a diapiric sense for rise of the forerange as more feeble top-to-southwest motion continued partly at a higher structural level. The brittle component of mylonitization increased upward in later-stage zones. Mylonites formed as deep as 8.7 to 10.1 km and as shallow as about 4.5 to 5.9 km. Total top-to-southwest shear was about 1.6 km. The main zone dips under the main range; the Windy Point zone that projects over it represents no more than 20% of the total mylonitic fabric.

STRUCTURE

Supracrustal rocks of the range dip northeast as a result of mid-Tertiary events. Thus the south end of the range (described first) represents the greatest crustal depths. The Forerange anticline exposes the deepest rocks, thought to represent depths of about 8.8 to 10.2 km in the mid-Tertiary (due to inflation by Eocene sills), but only about 4.8 to 6.2 km in the Laramide.

Post-mylonite faulting in the forerange (at about 21 Ma) was on the arcuate Sabino and Bear Canyon faults, and the Romero Pass fault zone. The forerange was uplifted as much as 1200 m on the latter, a reverse fault zone that everywhere separates the forerange from the main range. The Sabino and Bear Canyon faults rotate the block between them about an axis near Thimble Peak. The Catalina detachment fault cuts these structures. Thus detachment seems little-related to mylonitization in this area either in spatial locus or in precise time of formation.

In the range crest area, pre-Wilderness structures of Laramide age include recumbent sheath folds and above-described fabrics. These structures most commonly trend WNW, at a high angle to later mylonitic fabrics. The oldest generation of faults predates Leatherwood granodiorite. Among these structures north of the crest are the Edgar and Araster thrusts, which probably total well over a kilometer of offset. A family of small Eocene thrust faults occurs in the range-crest area also. Together with folding, these structures represent over 3 km of east-west shortening, apparently east side over west.

The Geesaman fault formed early in the Laramide (about 70 Ma) as a steep left-oblique fault with about 7 km of offset. It was subsequently folded by the Marble Peak syncline, such that its eastern end dips gently south. A rotated splay now is exposed as a window in Alder Creek. Leatherwood granodiorite intrudes the Geesaman fault plane, whereas the Marble Peak syncline is syn-Leatherwood.

The Marble Peak and Oro synclines and Lariat anticline and the associated Geesaman and Limb faults together represent SW-facing faulted monoclines of Laramide age, subsequently tilted in the mid-Tertiary. Coupled with NE-vergent reverse faults that bound the American Flag Formation lower member (fig. 6), these structures delineate a Laramide basement-cored uplift.

The Mogul fault, like the parallel Geesaman fault, began as a steep left-oblique fault with over 7 km offset, probably also in Laramide time but perhaps earlier. It was re-activated in the late Tertiary.

In the northern part of the study area, three generations of extensional faults have differentially rotated areal domains. The most severely rotated domain is the upper plate of the Cloudburst-Turtle detachment system or Tar Wash allochthon, limited to the area north of the Turtle fault. Later rotation on the San Manuel and related faults affected both hanging- and footwall domains, i.e. the whole immediate region.

Eocene leucogranites seem the most suitable contributors to a crustal root that may have controlled core-complex uplift, based on their distribution and thickness. Laramide crustal thickening is suggested by metamorphic fabrics and thrust faults, but where best documented contributes much less to a crustal root than intrusion does.

MINERAL RESOURCES

Mineral deposits of the study area include the Laramide San Manuel-Kalamazoo porphyry copper system and the mid-Tertiary Mammoth district of polymetallic veins toward the northern end of the study area, the American Flag district probably in part of pre-Laramide age south of Oracle, the Campo Bonito and Southern Belle districts of probable mid-Tertiary age south of and related to the Mogul fault, a newly recognized population of mid-Tertiary deposits along the Geesaman fault, and skarn deposits of Laramide age at Marble Peak. Laramide deposits contain mostly copper, and mid-Tertiary deposits contain mostly gold and tungsten. Laramide deposits are mostly intrusion-centered, whereas mid-Tertiary deposits are mostly fault-related, including detachment faults (Mammoth district) and transcurrent faults (Geesaman fault, Campo Bonito district). Potential for further deposits is related to the American Flag unconformity in the San Manuel area, the Cloudburst-Turtle fault system in the Mammoth district, the Campo Bonito Formation and bounding unconformities in the Southern Belle district, the Geesaman fault, pre-Leatherwood faults of the Marble Peak district, intruded Abrigo Formation north of the range crest, and the Catalina detachment fault. Deposits of marble and slate also occur.

DISCUSSION OF DEFORMATION-INTRUSION HISTORY

In this section all parts of the study area are discussed together in order of chronological development, in a more free-ranging mode that includes some speculation.

LARAMIDE

The study area contains two Laramide terranes. The southern terrane is metamorphosed and shows tectonic thickening by both penetrative deformation and thrusting, whereas the northern terrane instead shows a basement-cored uplift and a possible foreland basin, outlined by faulted monoclines and by high-angle reverse faults.

Differing intrusive domains correlate with these boundaries; that of the basement-cored uplift may be a laccolith. However, this intrusion cannot have driven basement uplift, as it is emplaced in overlying supracrustal rocks.

The contrast between these terranes might be thought to be a function of paleodepth in a northeast-dipping homoclinal sequence, but the paleodepths determined herein suggest that depth contrast was not sufficient to explain the contrast. Tectonic contrasts reflected in areal distribution are apparently more important.

In the southern terrane, extensive Laramide deformation and recrystallization of older rocks affected the range crest area and apparently extended southward. Metamorphism was intrusion-related, and produced regional greenschist assemblages to local high-grade skarn assemblages. Recumbent sheath folding was common in the range-crest area. Laramide fabrics and metamorphic assemblages to the south were extensively overprinted by younger events. Paleodepths of exposed rocks were as great as about 4.8 to 6.2 km; geothermal gradients must have been high.

Northward from the range crest area, Laramide ductile fabrics and metamorphism decrease rapidly past the Edgar thrust set. Laramide faults include the Geesaman and Mogul transcurrent oblique faults, and reverse faults near Nugget Canyon. Laramide folding included a SW-facing monocline. The monocline and reverse faults together define a Laramide basement-cored uplift at least 5 km across in the northern part of the range.

American Flag Formation serves as an intricate recorder of Laramide events near the ground surface on the northeastern margin of the basement-cored uplift. Eastward-directed reverse faulting, perhaps fold growth, and porphyry intrusion was followed by porphyry unroofing. A paleosurface below American Flag Formation and vent facies of older Rice Peak porphyry was at the level of Escabrosa Limestone in this area, but was on Oracle Granite in the San Manuel mine area, probably reflecting early- to pre-Laramide uplift north of the Mogul fault. Faulting synchronous with American Flag accumulation suggests a proto-San Pedro foreland(?) basin of Laramide age, and northward continuation of the formation suggests similar continuation of the basin margin.

Mostly-concordant Laramide intrusions inflated older sequences typically by about 300 m across the study area, but differ greatly in their effects. Two areas that were less intruded are southeast of Alder Creek and near Oracle. A semi-concordant Leatherwood granodiorite sheet intruded the deeper southern parts of the area, at depths between about 1.9 and 4.6 km. It produced extensive skarn and other high T/P metamorphism. At shallower levels to the north are the Rice Peak and San Manuel porphyries, both intruded partly discordantly and partly concordantly.

The areas occupied by these Laramide intrusions are almost mutually exclusive, and they record distinct conditions not only of depth but of intrusive, alteration, and mineralization character. For example, the older Rice Peak intrusion, coincident with the area of basement-cored uplift, is shaped like a laccolith (but with discordant arms), supplied little heat, and lacks much hydrothermal alteration or mineralization, whereas Leatherwood granodiorite to the south is a tabular intrusive sheet consistently discordant in detail, and was a prolific supplier of heat to form skarn deposits and high T/P metamorphism. A discordant San Manuel porphyry to the north drove a large hydrothermal alteration system and produced a large copper deposit.

The Geesaman and Mogul faults both originated rather early in the Laramide as steep left-oblique faults, each with several kilometers of displacement. The Geesaman was folded by the Marble Peak syncline, synchronous with Leatherwood intrusion, still in the Laramide.

Comparison of the 4.8 to 6.2 km of Laramide crustal section in the Catalinas with an almost equally thick pile of Laramide volcanics in the Tucson Mountains (Lipman, 1993) shows few similarities. Probably in the Laramide, the two areas were nearby but not superposed.

EOCENE

Intrusion of seven leucogranite sills, aggregating 3.9 km in thickness, greatly inflated the section and added significantly to crustal thickness in the southern Catalinas. Determination of the chronology of intrusion of these Eocene sills would have implications for a thermal history. Wilderness sill emplacement produced little ductile fabric except locally near the range crest, where some porphyroblasts and crenulation of older fabrics formed. However, the continuation of a compressive regime is shown by syn-Wilderness motion on the Bear Wallow thrust fault and by weak foliation in Spencer granite.

MID-TERTIARY

The study area in the mid-Tertiary was greatly extended at shallow- and mid-crustal levels, at two discrete times in each part of the area. These events may correlate between subareas. Plutons of mid-Tertiary age help the discrimination of extensional events in the range-crest area. Discussion is from deep to shallow features.

Mylonitization affected all the rocks of the Forerange and some of the main range. Mylonitization was most pervasive at paleodepths of 8.1 to 10.1 km, but some occurred at depths of only about 4.5 to 5.9 km. Pure shear was a component of early mylonitic deformation. The main mylonite belt dips under the main range, suggesting that mylonite continues under much of it. This belt consists of semi-ductile rocks overlying fully ductile rocks; it must originally have been planar, probably subhorizontal, but growth of the Forerange anticline stopped further mylonitization on the original locus. Reversal of movement direction on the northeast limb of the anticline gives the reactivated component of the main mylonite belt a diapiric sense of motion. The original top-to-southwest geometry of mylonitic slip was apparently preserved as late-stage mylonitization along the Windy Point zone at higher structural levels. However, the core-complex aspect of the geology of the Catalinas is a function mostly of the earlier stage of mylonitization and its folding.

The shallow dip of the upper intrusive contact of Catalina granite in the Mt. Lemmon area, and the presence of the large Knagge pluton probably of mid-Tertiary age at the eastern end of the range, suggest that mid-Tertiary plutons may underpin much of the range-crest area, but are exposed only at the ends. This may have thermal implications for core-complex uplift. The Catalina granite seems to cut the main, deeper mylonite zone but is locally cut by the younger, higher Windy Point zone. The Knagge granite, however, cuts the Windy Point zone.

Detachment faulting on the older Cloudburst-Turtle fault system in Tar Wash allochthon, and the younger San Manuel low-angle normal fault and related faults, led to deposition of 5.5 km of coarse clastics and volcanics in composite basins as older Cloudburst Formation and younger San Manuel Formation. These basins were the mid-Tertiary precursors to the San Pedro basin. Progressive crustal extension led to local progressive tilting of as much as 70°. Detachment faulting on the Stratton fault led to similar accumulation on the proto-San Pedro margin in an area that is now internal to the study area; it has probably been offset there by younger movement on the Mogul fault.

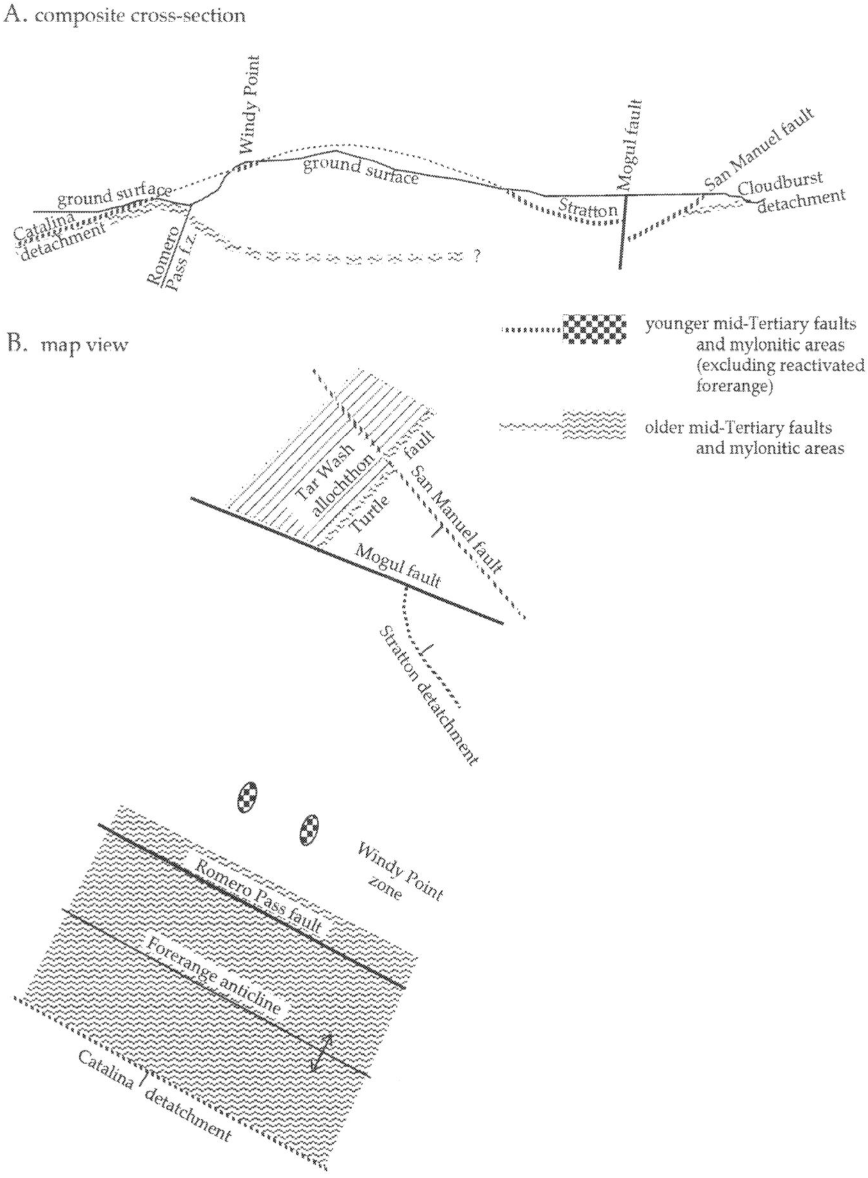

Figure 108 Diagrams of proposed correlation of mid-Tertiary extensional structures. A. Cross-sectional correlations; B. Map view.

On the southern margin of the study area, brittle faulting followed mylonitization, tilted one block, and raised the Forerange relative to the main range. Final movement on the Catalina detachment fault postdates these brittle faults, showing departure of detachment faulting from mylonitization not only in spatial locus but in time of development. Detachment faulting on the Catalina fault and its precursors led to deposition in two stages of up to 3.5 km of coarse clastics in proto-Tucson basins.

The correlation of detached areas and episodes is uncertain, and hence the original continuity and correlation of detachment faults over the Catalinas is problematic. For example, the tilt block associated with movement on the San Manuel fault apparently extends to the Mogul and Stratton faults. If so, steep dips of the Stratton unit above the Stratton fault could represent this tilt block, further tilted by core-complex uplift of the Catalinas (fig. 108), and the Stratton fault could correlate with the San Manuel generation of low-angle normal faults. This correlation would in turn imply extension of the Stratton detachment fault over the range to correlate with the Catalina detachment fault and (or) the Windy Point mylonite zone. The lower part of Tinaja beds would probably correspond to this stage of detachment faulting. This suggestion is broadly consistent with age constraints on these features, as they all may postdate 26 Ma; it is also consistent with original continuity of a Catalina-San Pedro-Stratton detachment system around the east end of the Catalina and Rincon Mountains (fig. 6).

Older mid-Tertiary extensional structures that may correlate are the first period of movement on the main mylonite belt and the Cloudburst-Turtle detachment system. This correlation makes sense in terms of relative timing, paleodepth trends, and cross-sectional relations to the main range (fig. 108), but makes less sense in map view because the main range would seem to be outside the Tar Wash allochthon. The main mylonite zone may be a regional feature only locally tapped by detachment faults. Cloudburst and Pantano Formations might correlate in part; this group of features all shows a pre-26 Ma history. The implied correspondence of Pantano Formation with the main mylonite belt is problematic, as that belt lacked a detachment fault at presently exposed paleodepths. However, in any scenario the distribution of Pantano Formation is an artifact of later-stage detachment.

Comparison of the distributions of the mid-Tertiary mylonitic core complex in the southernmost parts of the range with inherited Laramide features shows poor correspondence with Laramide thrust and reverse faults, monoclinal fold, and basement-cored uplift, but possibly good correspondence with a zone of Laramide metamorphic fabrics. A more certain and likely correspondence is with the zone of thickening by Eocene leucogranite sills. This may have produced a crustal root suitable for isostatic core-complex uplift when truncated by detachment (e.g. Myers and Beck, 1994).

Mineralization of mid-Tertiary age is along faults, and includes detachment-related deposits (Mammoth district) and deposits along the older Mogul and Geesaman faults, some related to Catalina granite. Mineralization was largely of gold and tungsten.

LATE TERTIARY

Among the younger faults of the area (pl. 1) are basin-range faults that arc from a NW-SE strike toward the north end of the study area, through a north-south strike toward the south. They enter the range as north-south faults, but continue their arc to become NE-SW-striking steep faults, mostly down to the southeast. The described arc is somewhat similar to but gentler than that of the older Sabino and Bear Canyon faults. It is more similar to that of the Pirate fault at the west end of the range (fig. 6), but with the opposite sense of dip-slip movement. The intermediate area is therefore a horst. In this sense the topography of the Catalinas is enhanced by parallel arcuate basin-range faults. However, the magnitude of uplift attributable to these faults at the east end of the range is only about 100 m toward the south end of the study area.

CONCLUSIONS ON REGIONAL CONTEXT

The great north-south extent of the study area has permitted the determination of spatial variations in both Laramide and mid-Tertiary histories, which in turn allows rigorous analysis of separate components of this history and permits a glimpse at the place of the study area in regional tectonic patterns.

Laramide crustal behavior differs in the northern and southern parts of the study area. The northern terrane is characterized by a basement-cored uplift, a possible foreland basin, faulted monoclines, high-angle reverse faults, and a possible laccolith. The southern terrane is characterized by penetrative tectonic fabrics, thrust faults, and pluton-centered metamorphism. Thus in the Laramide the study area formed a sort of "transition zone" between crust that behaved like the Colorado Plateau and more highly strained crust to the south and west (cf. Krantz, 1989) that is more comparable with that of south-central Arizona, in both orogenic character (Haxel and others, 1984) and Laramide timing (Wright and Haxel (1982).

These transitions of Laramide features in the study area are somewhat telescoped by thrust faults and juxtaposed laterally by the left-lateral strike-slip components of the Geesaman and Mogul faults. When restored, the NNE-SSW elongation of the study area becomes NE-SW, consistent with the proposed transition from Plateau to orogen.

In the mid-Tertiary, the southern end of the study area acquired the classic characteristics of a metamorphic core complex. The apparently close correspondence of detachment and mylonitization in this well-studied part of the area led to a concept of earlier, deeper shear along the same plane that later became a detachment at shallower crustal levels. This study shows great departures of detachment and main-stage mylonitization both spatially and temporally. Though both are mid-Tertiary in the study area, the pattern is not unlike that of detachment and older mylonite in some mountain ranges of southeastern California (cf. Davis and others, 1982; John and Mukasa, 1990).

Probable correlation of a post-26 Ma (San Manuel-Stratton-Windy Point-Catalina) generation of structures and deposits over the range is gratifying but not surprising. Correlation of an older pre-26 Ma Cloudburst-main mylonite generation of features under the main range is problematic, requiring some unconstrained shifts in extensional domain boundaries. An alternative approach is to look at the main mylonitic belt as a domed exposure of regional sub-horizontal ductile deformation, found to be laterally extensive to the northeast (Kruger and Johnson, 1994). This view is consistent with its pure shear component and the boundary with fully-ductile rocks at its base.

Two causes of core-complex uplift proposed elsewhere are both permitted by relations in the study area. Mid-Tertiary intrusives are available to contribute to heat-driven uplift, but are not coincident spatially with the locus of uplift along the Forerange anticline. An earlier crustal root may have been truncated by detachment, allowing isostatic uplift; geophysical evidence shows a remnant root still present (Myers and Beck, 1994). Both Laramide crustal thickening and Eocene leucogranite sill emplacement show appropriate distributions to contribute to a crustal root; the sills contribute about 4 km whereas crustal thickening contributes only about 1 km.

The concept of a metamorphic core complex generally includes a metamorphic core and a detachment rim associated with mylonitic rocks, with some concentric distribution due to uplift. The Catalinas in this framework constitute a very lopsided core complex. That is, from the southern part that has the characteristic mylonitic rocks exposed in an anticline, the range continues northward for 30 km as a tilted block including other metamorphic rocks, above the main mylonite zone but below the main detachment fault. Thus the study area poses a dilemma — how does one deal with a complex where "lower plate" mylonites have one distribution, older metamorphic rocks have a mostly-separate distribution, and the geometry of detachment faults yet another? How independent can these variables be and leave the term "metamorphic core complex" a useful description?

Porphyry Geochronology and Geochemistry

Daniel M. Unruh

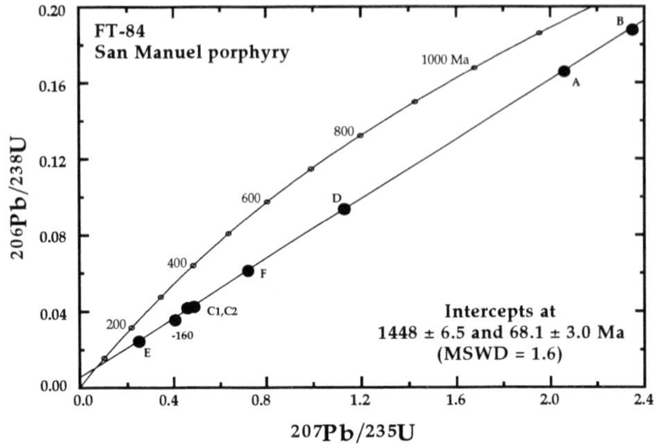

Figure 109 U-Pb concordia diagram for zircon separates from San Manuel porphyry sample FT 84. Individual zircon fractions are defined in table 11. Uncertainties are generally smaller than the symbols shown. Ages and uncertainties in this and subsequent figures 110 to 112 are calculated in the manner of Ludwig (1984).

PURPOSE

The age of San Manuel porphyry prior to this study was known only from K-Ar dates, and the age of related Rice Peak porphyry had not been determined by any radiometric method. Because the intrusion of these units provides fixed points in the tectonic evolution of the northern Santa Catalina Mountains, precise ages were needed.

Force (this volume) finds that each of these porphyry units consists of two types, although the apparently younger San Manuel type A porphyry is not exposed. Accordingly, four samples of three units were collected for U-Pb analyses of zircon; two are now thought to be San Manuel type B porphyry (the identity of one being unclear at the time of sampling), and the other two are of each type of Rice Peak porphyry. Only the San Manuel porphyry samples yielded reliable ages, but the data are presented for all four samples (tables 7, 11–14).

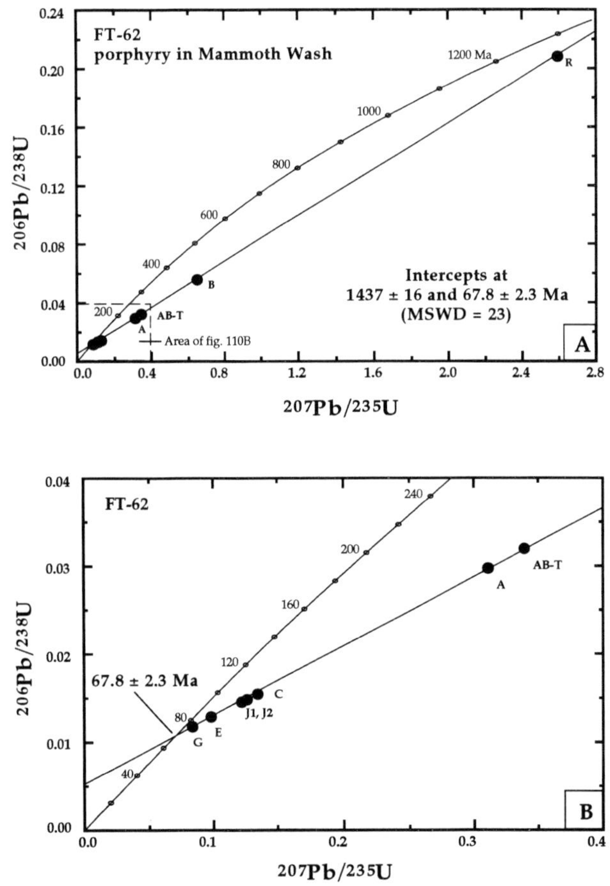

Figure 110 U-Pb concordia diagram for zircon separates from San Manuel porphyry sample FT 62. A. Whole-chord; B. Detail of lower left corner of 110A.

SAN MANUEL PORPHYRY

San Manuel porphyry was previously dated by S. C. Creasey using K/Ar on biotite from specimens from the San Manuel mine. Creasey (oral commun., 1991) is of the opinion that the biotite was primary, but Reynolds and others (1986) have interpreted it as hydrothermal. Two samples gave ages of 65.4 and 69.0 Ma (S. C. Creasey, written commun., 1991).

San Manuel type B porphyry was sampled in two places for this study. The first locality (loc. 72, sample FT 84) in the Purcell window area (pl. 1) is the freshest known exposure of San Manuel porphyry, and is Jon Hagstrum's site 1 (in Force and others, 1995) for paleomagnetic analysis. The host rock for the porphyry is Oracle Granite (1400-1440 Ma). The porphyry body sampled is the thick northwest-trending dike (pl. 1), shown by Lowell (1968) to be a Christmas-tree-shaped body dipping northeast. The collection site is about 850 m (2800 ft) above the Kalamazoo segment of the San Manuel ore body; the sample shows slight deuteric alter-ation. The assignment as type B porphyry is based on the petrography of the sample and the orientation of the dike.

The other sample (FT 62, loc. 55) was collected in the south fork of Mammoth Wash, in a narrow horst between the San Manuel and Cholla faults (pl. 1). Porphyry occurs as a NNW-trending dike; it is not everywhere fault-bounded as shown by Creasey (1965). The context of this sample was unclear at the time of collection, because the porphyry body sampled is intrusive mostly into conglomerate and intermediate-composition volcanic rocks mapped as Cloudburst Formation (late Oligocene to early Miocene) by Creasey (1965). Based upon the U-Pb zircon date reported here, petrographic data, and the trend of the dike, Force and others (1995) regard the porphyry represented by FT 62 as San Manuel type B porphyry, and the intruded host rocks as probable American Flag Formation of late Cretaceous age. Near the sample site, this porphyry also intrudes Oracle Granite. The sampled exposure of porphyry is slightly weathered.

Zircon Morphology

Zircons from San Manuel porphyry are typically clear, euhedral, colorless to pale brown, and somewhat elongate with complex, steeply pyramidal terminations. However, nearly-equant crystals as well as acicular crystals were also found in the finest size-fractions. Although no obvious zoning was observed in the zircons, it was found that they contain from 1 to 80 percent "inherited" Pb (figs. 109, 110). In other words, these zircons contain cores of xenocrystic zircon that were not isotopically equilibrated with the magma at the time of emplacement, a commonly observed

Table 11 U-Pb data for zircon separates from San Manuel porphyry samples FT 84 and FT 62.

Fraction*	Wt. (mg)	Pb ppm	U ppm	$^{204}Pb/^{206}Pb$†	$^{206}Pb/^{238}U$‡	$^{207}Pb/^{235}U$‡	$^{207}Pb/^{206}Pb$‡
FT 84							
A	0.299	14.65 ± 0.34	70.6 ± 0.2	0.00101	0.16606 ± 0.00102	2.0585 ± 0.0134	0.08991 ± 0.00030
B	1.110	13.88 ± 0.18	59.5 ± 0.2	0.00156	0.18804 ± 0.00084	2.3498 ± 0.0123	0.09064 ± 0.00026
C1	0.606	5.83 ± 0.30	122.2 ± 0.4	0.00053	0.04126 ± 0.00049	0.4643 ± 0.0059	0.08161 ± 0.00044
C2	1.021	4.20 ± 0.06	82.8 ± 0.2	0.00120	0.04185 ± 0.00022	0.4728 ± 0.0161	0.08192 ± 0.00253
D	0.788	10.63 ± 0.06	100.3 ± 0.3	0.00038	0.09372 ± 0.00036	1.1345 ± 0.0072	0.08779 ± 0.00042
E	0.565	3.86 ± 0.08	131.3 ± 0.4	0.00102	0.02457 ± 0.00015	0.2496 ± 0.0099	0.07369 ± 0.00027
F	0.690	8.04 ± 0.06	117.1 ± 0.4	≤0.00006	0.06180 ± 0.00023	0.7243 ± 0.0031	0.08500 ± 0.00019
-160	1.987	4.69 ± 0.02	111.9 ± 0.5	0.00086	0.03635 ± 0.00019	0.3996 ± 0.0028	0.07973 ± 0.00034
FT 62							
A	0.759	7.15 ± 0.02	197.8 ± 1.4	0.00177	0.02966 ± 0.00031	0.3106 ± 0.0034	0.07596 ± 0.00019
B	0.860	10.80 ± 0.02	166.0 ± 0.5	0.00097	0.05585 ± 0.00038	0.6483 ± 0.0045	0.08418 ± 0.00012
AB-T	1.154	7.18 ± 0.01	202.2 ± 0.7	0.00056	0.03205 ± 0.00017	0.3399 ± 0.0019	0.07691 ± 0.00013
C	1.117	4.25 ± 0.01	244.4 ± 0.7	0.00083	0.01551 ± 0.00006	0.1347 ± 0.0017	0.06299 ± 0.00069
E	0.676	3.61 ± 0.02	264.7 ± 0.9	0.00031	0.01284 ± 0.00022	0.0981 ± 0.0026	0.05542 ± 0.00071
G	0.280	4.08 ± 0.03	308.3 ± 1.0	0.00090	0.01172 ± 0.00007	0.0837 ± 0.0008	0.05175 ± 0.00035
J1	[0.11]§	4.84 ± 0.10	309.2 ± 1.0	0.00016	0.01488 ± 0.00011	0.1258 ± 0.0011	0.06128 ± 0.00029
J2	[0.15]§	3.92 ± 0.06	239.1 ± 0.7	0.00116	0.01474 ± 0.00010	0.1228 ± 0.0013	0.06045 ± 0.00039
R	[0.07]§	104.40 ± 0.20	463.1 ± 1.4	0.00048	0.20800 ± 0.00119	2.5962 ± 0.0165	0.09053 ± 0.00023

* Fractions

A, +100mesh, clear, euhedral, and elongate
B, +100 mesh, clear, euhedral, nearly equant
AB-T, broken tips of +100 mesh crystals
C, -100, +130 mesh, same as A
D, -100, +130 mesh, same as B
E, -130, +160 mesh, same as A
F, -130, +160 mesh, same as B
G, -160, +270 mesh, same as A
J, -270 mesh, same as A
R, pink to red-brown slightly etched and metamict, xenocrysts
-160, -160 mesh, bulk analysis

† Corrected for Pb blanks of 0.02 to 0.06 ng and mass fractionation of 0.11±0.03 percent per mass during mass spectrometry

‡ Corrected for an initial Pb isotopic composition of 206Pb/204Pb = 17.30, 207Pb/204Pb = 15.44

§ Sample weights in brackets are estimated

phenomenon (e.g., Gebauer and Grunenfelder, 1979). In general, the smallest acicular zircons contain the smallest proportion of xenocrystic cores.

In FT 62 rare red to red-brown, clear to moderately metamict, and slightly rounded zircon crystals were observed. The simple prismatic morphology of these crystals, as well as their distinct color suggest that these crystals represent xenocrysts without overgrowths. This observation is confirmed by the U-Pb data presented below.

Analyses of FT 84

U-Pb analyses of eight zircon fractions from sample FT 84 are summarized in table 11 and figure 109. In the U-Pb evolution diagram the data define a linear array that intersects the concordia curve at 68.1 ±3.0 Ma and 1448.0 ± 6.5 Ma. The lower intercept is in good agreement with the two K-Ar ages previously mentioned (S. C. Creasey, written commun., 1991) and probably represents the best value for the time of emplacement of San Manuel type B porphyry.

The upper intercept is in good agreement, though perhaps slightly older, than published ages for Oracle Granite summarized by Reynolds and others (1986; 1400–1430 Ma). Consequently, the xenocrystic zircon cores in San Manuel porphyry appear to have been derived from the Oracle Granite host rocks. The relative positions of the data points along the chord reflect 6 to 74 percent inherited radiogenic Pb from xenocrystic zircon cores. The agreement between the upper intercept and the age of Oracle Granite, the relatively small amount of excess scatter in the data (mean square weighted deviation, MSWD = 1.6), and the relatively low U contents of the zircon fractions all suggest that the 68.1 Ma age is probably very close to the true age.

Sample FT 84 was also subjected to Rb-Sr and Sm-Nd analyses (table 12). Calculated initial $^{87}Sr/^{86}Sr$ values from whole-rock, plagioclase, and apatite data are somewhat different, probably as a result of weathering or hydrothermal alteration. The lowest initial $^{87}Sr/^{86}Sr$ value, 0.70527 in apatite, probably represents the most reliable magmatic value.

The calculated initial $^{143}Nd/^{144}Nd$ value of 0.512282 for acid washed plagioclase (table 12, eNd =

Table 12 Rb, Sr, Sm, and Nd concentrations and Sr and Nd isotopic compositions for mineral separates from San Manuel and Rice Peak porphyries.*

Fraction†	Rb ppm	Sr ppm	Sm ppm	Nd ppm	$^{87}Rb/^{86}Sr$	$^{87}Sr/^{86}Sr$	$^{87}Sr/^{86}Sr_I$‡	$^{147}Sm/^{144}Nd$	$^{143}Nd/^{144}Nd$	$^{143}Nd/^{144}Nd_I$‡	εNd_I‡
FT 84											
WR	53.5	717.4			0.216±5	0.705982±22	0.705767±23				
Apatite	15.9	983.4			0.0468±5	0.705313±39	0.705266±39				
Plag-1A	24.8	1675.0			0.0429±6	0.705615±25	0.705572±25				
Plag-1B	30.2	1900.0			0.0460±4	0.705643±25	0.705597±25				
Plag-2B	25.1	1820.1	0.277	2.551	0.0400±6	0.705572±21	0.705532±21	0.06568±17	0.512312±17	0.512282±19	-5.2
FT 62											
Apatite	1.71	1119.7			0.00442±5	0.705298±29	0.705294±25				
FT BH											
WR	77.2	275.9			0.810±19	0.712297±21	0.711492±28				
Apatite-1	9.8	724.0			0.0390±5	0.708831±20	0.708792±20				
Apatite-2	5.8	750.6			0.0225±2	0.708727±34	0.708705±34				
Plag-1A	62.1	487.6			0.369± 8	0.711796±20	0.711430±22				
Plag-1B	69.6	501.4			0.402±12	0.711784±21	0.711384±24				
Plag-2B	66.0	545.1	0.214	1.184	0.350±11	0.711709±42	0.711361±43	0.1094±19	0.512150±25	0.512100±25	-8.7
FT 820											
WR	55.8	780.9			0.207±5	0.709168±21	0.708963±22				
Apatite	1.2	513.1			0.0067±1	0.709714±25	0.709707±25				
Plag-1A	39.1	796.8			0.142±3	0.708870±21	0.708729±21				
Plag-1B	38.0	765.6			0.144±4	0.708887±17	0.708744±17				
Plag-2B	36.0	780.7	0.134	1.161	0.133±4	0.708807±21	0.708674±21	0.0679±15	0.512259±29	0.512227±29	-6.2

* Uncertainties correspond to last significant figures.

† Fractions: WR = whole-rock; Plag-1, leached with 2N HCl; Plag-2, leached with 1N HCl + 2N HF at 60 degrees C.

‡ Corrected for an assumed age of 68 Ma. Uncertainties do not include the uncertainty in the age.

-5.2), is only slightly higher than those reported by Farmer and DePaolo (1987) for altered samples from the San Manuel porphyry (eNd = -5.6 to -6.6), even though the initial $^{87}Sr/^{86}Sr$ value of the least-altered sample analyzed by them was considerably higher (0.7088) than that of FT 84. The results presented here support the conclusion of Farmer and DePaolo (1987), that Nd isotopic values are considerably less affected than Sr isotopic values by exchange with hydrothermal fluids, and the initial $^{143}Nd/^{144}Nd$ value of 0.512282 obtained for FT 84 probably reflects the magmatic value at the level of emplacement.

U-Th-Pb data for three fractions of FT 84 (table 13) also show small but distinct differences in the apparent initial Pb isotopic ratios among the fractions, with acid-leached plagioclase having the lowest values and the groundmass separate the highest. The U-Th-Pb systems are subject to the same post-emplacement processes that may have affected the Rb-Sr data presented above. I consider the values for acid-leached plagioclase, $^{206}Pb/^{204}Pb$ = 17.304, $^{207}Pb/^{204}Pb$ = 15.442, $^{208}Pb/^{204}Pb$ = 37.320, to be most representative of the magmatic values.

Analyses of FT 62

U-Pb data for nine zircon fractions from FT 62 are shown in table 11 and figure 110. Zircons from FT 62 have much lower amounts of inherited radiogenic Pb than those from FT 84. Zircon fractions from FT 62, with the exception of fraction "R", contain 0.4 to 18 percent inherited Pb and five of the nine fractions have less than 2 percent. Fraction "R" probably represents pure xenocrystic zircon that suffered about 18 percent Pb loss relative to U during incorporation into the magma.

The U-Pb data define a linear array that intersects the concordia curve at 67.8 ± 2.3 Ma and 1437 ± 16 Ma (fig. 110). As with FT 84, the 67.8 Ma age reflects the time of emplacement of the porphyry and the 1437 Ma age reflects that of Oracle Granite. The data show significant scatter about the best fit line (MSWD of 23). The scatter results from variable amounts of post-emplacement Pb loss resulting from either weathering or from radiation damage to zircon in FT 62 as a consequence of the higher U contents in zircon from this sample (table 11).

The chord defined by the data for FT 62 is identical, within analytical uncertainty, to that defined by the data for FT 84. There is no evidence to suggest that this porphyry body is significantly younger than San Manuel porphyry, and the intruded sedimentary host rocks cannot be Cloudburst Formation as currently defined (see the discussion by Force elsewhere in this volume).

Because the porphyry represented by FT 62 is somewhat weathered, few trace element data were obtained for this sample. Rb-Sr data for an apatite separate (table 11) from FT 62 yield an apparent initial

Table 13 U, Th, and Pb concentrations and Pb isotopic compositions of mineral separates from San Manuel and Rice Peak porphyries.

Sample*	U ppm	Th ppm	Pb ppm	$^{238}U/^{204}Pb$	$^{206}Pb/^{204}Pb$	$^{207}Pb/^{206}Pb$	$^{208}Pb/^{206}Pb$	Initial† $^{206}Pb/^{204}Pb$	Initial† $^{207}Pb/^{204}Pb$	Initial† $^{208}Pb/^{204}Pb$
FT 84										
GM-1	1.27	4.53	18.25	4.320±.041	17.501±.016	0.88363±.00054	2.1449±.0016	17.454±.016	15.463±.015	37.485±.047
GM-2	1.38	4.10	18.49	4.619±.095	17.501±.015	0.88283±.00031	2.1436±.0015	17.450±.016	15.449±.017	37.465±.052
Apatite	4.30	15.35	8.92	29.96±.10	17.690±.011	0.87393±.00027	2.1335±.0013	17.362±.011	15.444±.014	37.372±.047
Plag-1	0.055	0.111	7.04	0.483±.002	17.351±.011	0.89044±.00026	2.1537±.0013	17.346±.011	15.450±.014	37.365±.052
Plag-2	0.026	0.060	6.78	0.233±.001	17.307±.011	0.89225±.00029	2.1565±.0013	17.304±.011	15.442±.014	37.320±.046
FT BH										
GM	0.91	3.86	7.56	7.588±.043	18.243±.014	0.85153±.00026	2.0941±.0013	18.160±.014	15.530±.016	38.091±.049
Apatite	3.86	—	10.45	23.83±.11	19.206±.013	0.81559±.00025	2.0227±.0012	18.945±.013	15.653±.015	—
Plag-1	0.046	0.30	6.11	0.477±.003	18.146±.014	0.85518±.00026	2.1000±.0013	18.141±.014	15.518±.016	38.096±.048
Plag-2	0.029	0.22	5.83	0.308±.003	18.140±.014	0.85594±.00027	2.0998±.0013	18.137±.014	15.527±.016	38.082±.048
FT 820										
GM	0.92	3.99	11.23	5.040±.026	17.493±.013	0.88392±.00028	2.1588±.0013	17.438±.013	15.459±.016	37.687±.049
Apatite	9.07	12.71	4.82	124.1±1.3	21.004±.016	0.74951±.00030	1.8348±.0012	19.649±.021	15.677±.015	37.934±.048
Plag-1	0.055	0.33	1.69	2.032±.013	17.519±.014	0.88336±.00036	2.1593±.0014	17.497±.014	15.475±.015	37.786±.048
Plag-2	0.016	0.13	0.96	1.053±.005	17.480±.011	0.88475±.00028	2.1606±.0013	17.469±.011	15.465±.014	37.736±.046

* Mineral separates: GM = groundmass, Plag-1= plagioclase leached with 2N HCl, Plag-2 = plagioclase leached with hot 1N HCl+2N HF.

† Corrected for an age of 68 Ma. Uncertainties do not include the uncertainty in the age.

$^{87}Sr/^{86}Sr$ of 0.70529 ± 0.00003. This value is identical, within analytical uncertainty, to that obtained from the apatite separate of FT 84, reinforcing the correlation of these two porphyry bodies.

Petrogenetic Conclusions

Hydrothermal alteration may facilitate exchange of Sr and Pb between intrusions and host-rocks. In their isotopic study of the San Manuel ore body, Farmer and DePaolo (1987) found that apparent initial $^{87}Sr/^{86}Sr$ increased from 0.7084 in the sericitized zone of San Manuel porphyry to 0.7144 in the potassic alteration zone. They interpreted this variation to reflect exchange of Sr, via hydrothermal fluids, between porphyry and host Oracle Granite ($^{87}Sr/^{86}Sr$ = 0.7815 at 70 Ma).

Based on their Nd and Sr study, Farmer and DePaolo also suggested that the porphyry formed from a mantle-derived magma that had interacted with substantial amounts of lower-crustal material with a moderately low $^{87}Sr/^{86}Sr$. The Sr, Nd, and Pb isotopic data presented here support this interpretation. The initial Pb isotopic ratios plot within the low-$^{208}Pb/^{204}Pb$ portion of the lower-crust field of Zartman and Doe (1981), and the combined Pb, Sr, and Nd data (including that of Farmer and DePaolo) suggest that the isotopic characteristics of this porphyry were derived primarily from the lower crust, with some mantle contribution (e.g. Zartman, 1984) as well as a contribution from the upper crust (Oracle Granite) as indicated by the presence of xenocrystic zircon.

Owing to magmatic and possibly hydrothermal interaction with host Oracle Granite, the initial Sr and Pb isotopic values reported here are probably slightly higher, and the initial Nd value slightly lower than those in the parental magma at the site of melt generation.

RICE PEAK PORPHYRY

One sample of each of the two types of Rice Peak porphyry was collected. The first sample, FT BH (loc. 67) was collected along Oracle Ridge north of Rice Peak (pl. 1). It represents the laccolithic part of the older variety of Rice Peak porphyry and is intrusive into the upper member of the Pioneer Formation. The sample is a foliated plagioclase-amphibole-biotite porphyry. Epidote-chlorite-calcite alteration typical of the immediate area is well-represented in this sample.

Sample FT 820 (loc. 52) is a trachytic-textured plagioclase-amphibole-biotite porphyry from a sill intruded along the base of the middle member of the American Flag Formation (pls. 1, 2A). The sampled rock is assigned to the younger of the two varieties of Rice Peak porphyry. This rock is altered to a similar assemblage, but to a lesser extent than FT BH, and shows no distinct foliation. Small (<1 mm) euhedral pyrite crystals are present, and apatite has an etched and rounded appearance.

Zircon Morphology

Zircons separated from Rice Peak porphyry appear similar to those from San Manuel porphyry; the reddish xenocrystic population is present in both samples. In sample FT 820 a few clear, euhedral, and nearly equant pale pink crystals were observed. U-Pb

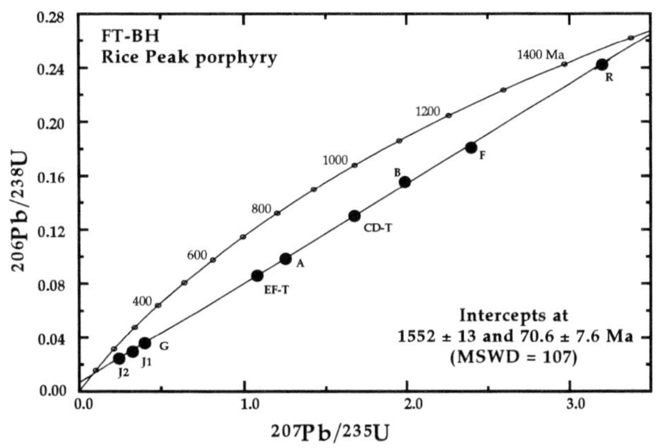

Figure 111 U-Pb concordia diagram for zircon separates from Rice Peak porphyry (older variety) sample FT BH. Individual zircon fractions are defined in table 14.

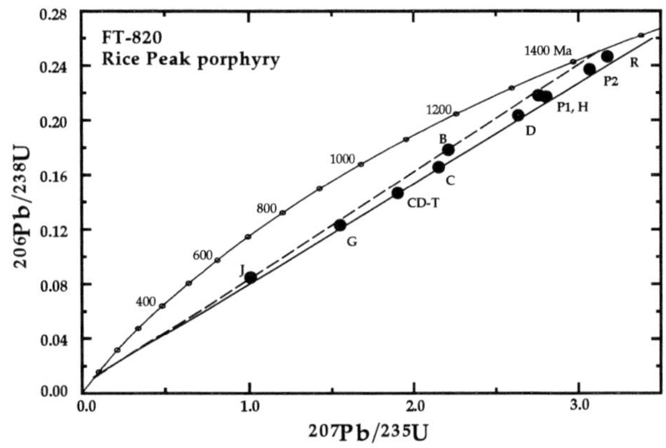

Figure 112 U-Pb concordia diagram for zircon separates from Rice Peak porphyry (younger variety) sample FT 820. Individual zircon samples are defined in table 14. Reference chords are for San Manuel porphyry samples (broken line), and the older variety of Rice Peak porphyry (solid line).

analyses suggest that these crystals are also xenocrystic, with perhaps thin overgrowths.

Analyses

U-Pb analyses of zircon from these two samples are summarized in table 14 and figures 111 and 112. The analyses failed to yield reliable age information. Zircons contain a large inherited component —all fractions of FT 820 contain over 40 percent xenocrystic Pb, and most fractions of FT BH contain over 30 percent inherited Pb. In addition the xenocrystic zircon was apparently derived from more than one source; ^{207}Pb/^{206}Pb model ages of fractions with the largest inherited component (>60 percent) range from about 1480 Ma to 1580 Ma, so the U-Pb data exhibit considerable scatter when plotted on a concordia diagram (MSWD>100).

A best-fit line through the data for FT-BH (fig. 111 intersects the concordia curve at 1552 ± 13 Ma and 70.6 ± 7.6 Ma. The data exhibit considerable scatter as exemplified by the MSWD = 107. Furthermore, the apparent 1552 Ma upper intercept corresponds to no known igneous event in southern Arizona. Consequently, the signifcance of the chord in terms of true age information is open to question. In any case the 70.6 ± 7.6 Ma lower intercept age is not precise enough to make a meaningful comparison with those of San Manuel porphyry samples.

U-Pb data for zircon separates from FT-820 (fig. 112) exhibit even more scatter than those of FT-BH. The broken line in figure 112 is a 68 Ma–1448 Ma reference chord obtained for San Manuel porphyry.

Table 14 U-Pb data for zircon separates from Rice Peak porphyry samples FT BH and FT 820.

FRACTION*	Wt. (mg)	Pb ppm	U ppm	^{204}Pb/^{206}Pb†	^{206}Pb/^{238}U‡	^{207}Pb/^{235}U‡	^{207}Pb/^{206}Pb‡
FT BH							
A	0.481	20.77 ± 0.03	192.7 ± 0.6	0.00040	0.09815 ± 0.00095	1.2504 ± 0.0129	0.09240 ± 0.00031
B	0.557	19.45 ± 0.02	108.0 ± 0.3	0.00019	0.15616 ± 0.00103	1.9884 ± 0.0179	0.09234 ± 0.00051
CD-T	0.258	24.15 ± 0.07	177.6 ± 0.5	0.00009	0.13005 ± 0.00076	1.6773 ± 0.0099	0.09355 ± 0.00008
F	0.510	37.36 ± 0.03	198.2 ± 0.6	0.00005	0.18127 ± 0.00118	2.3934 ± 0.0157	0.09576 ± 0.00008
EF-T	0.621	14.90 ± 0.01	165.2 ± 0.5	0.00017	0.08604 ± 0.00049	1.0823 ± 0.0062	0.09123 ± 0.00007
G	0.615	7.11 ± 0.01	179.2 ± 0.6	0.00035	0.03581 ± 0.00020	0.4107 ± 0.0023	0.08317 ± 0.00010
J1	0.120	5.82 ± 0.12	175.8 ± 0.6	0.00074	0.02954 ± 0.00019	0.3301 ± 0.0035	0.08106 ± 0.00062
J2	0.085	5.29 ± 0.12	146.6 ± 0.5	0.00584	0.02439 ± 0.00030	0.2474 ± 0.0048	0.07356 ± 0.00062
R	0.105	171.40 ± 0.20	707.0 ± 2.6	0.00009	0.24199 ± 0.00280	3.2029 ± 0.0372	0.09599 ± 0.00010
FT 820							
B	0.233	31.24 ± 0.03	158.7 ± 0.5	0.00060	0.1787 ± 0.0007	2.214 ± 0.010	0.08985 ± 0.00012
C	0.329	35.60 ± 0.03	203.6 ± 0.8	0.00019	0.1658 ± 0.0010	2.152 ± 0.014	0.09413 ± 0.00014
D	0.596	42.44 ± 0.02	190.6 ± 0.6	0.00005	0.2035 ± 0.0010	2.641 ± 0.013	0.09416 ± 0.00008
CD-T	0.294	25.76 ± 0.03	163.7 ± 0.5	0.00035	0.1477 ± 0.0011	1.916 ± 0.015	0.09470 ± 0.00009
G	[0.02]§	26.50 ± 0.43	203.3 ± 0.8	_0.00003	0.1230 ± 0.0009	1.552 ± 0.013	0.09153 ± 0.00031
H	[0.04]§	49.53 ± 0.23	211.7 ± 0.8	_0.00003	0.2199 ± 0.0016	2.820 ± 0.039	0.09300 ± 0.00098
J	[0.03]§	14.42 ± 0.08	167.7 ± 0.6	_0.00001	0.0854 ± 0.0005	1.013 ± 0.011	0.08605 ± 0.00082
P1	0.139	66.70 ± 0.07	294.3 ± 0.9	0.00009	0.2198 ± 0.0016	2.776 ± 0.020	0.09161 ± 0.00007
P2	[0.06]§	66.30 ± 0.17	261.9 ± 0.9	0.00051	0.2395 ± 0.0020	3.095 ± 0.035	0.09375 ± 0.00070
R	0.262	87.12 ± 0.05	339.8 ± 1.1	0.00014	0.2496 ± 0.0016	3.202 ± 0.020	0.09306 ± 0.00011

* Fractions

A, +100mesh, clear, euhedral, and elongate
B, +100 mesh, clear, euhedral, nearly equant
C, -100, +130 mesh, same as A
D, -100, +130 mesh, same as B
CD-T, broken tips of -100, +130 mesh crystals
F, -130, +160 mesh, same as B
EF-T, broken tips of -130, +160 mesh crystals
G, -160, +270 mesh, same as A
H, -160, +270 mesh, same as B
J, -270 mesh, same as A
P, pink euhedral to slightly rounded
R, pink to red-brown slightly etched and metamict, xenocrysts

† Corrected for Pb blanks of 0.02 to 0.06 ng and mass fractionation of 0.11±0.03 percent per mass during mass spectrometry

‡ Corrected for an initial Pb isotopic composition of 206Pb/204Pb = 17.50, 207Pb/204Pb = 15.48(FT 820) and 206Pb/204Pb = 18.14, 207Pb/204Pb = 15.52 (FT BH)

§ Sample weights in brackets are estimated

The lower, solid line is the 71Ma–1552 Ma chord obtained for FT-BH data. Data for FT-820 are scattered between these two chords and do not provide any meaningful age information.

Alteration of Rice Peak porphyry samples precludes extensive conclusions from trace-element and isotopic data. Rb-Sr and U-Th-Pb analyses of plagioclase and apatite separates from the two Rice Peak porphyry samples did not yield consistent or reliable results (tables 12 and 13). Apparent initial $^{87}Sr/^{86}Sr$ values of 0.7087 and 0.7114 were obtained for apatite and plagioclase, respectively, from FT BH; values of 0.7097 and 0.7087 were obtained from these two respective separates from FT 820 (table 12). U-Th-Pb data yielded similarly disparate results (table 13). Apatite separates show much more radiogenic apparent initial Pb ($^{206}Pb/^{204}Pb$ = 18.9 and 19.6 for FT BH and FT 820, respectively) than plagioclase separates ($^{206}Pb/^{204}Pb$ = 18.14 and 17.47 for FT BH and FT 820, respectively, table 13). Consequently, although the lower apparent initial $^{87}Sr/^{86}Sr$ values of 0.7087 may be taken as "best-estimates" of the magmatic values for both samples, the true magmatic values may have been much lower.

Initial $^{143}Nd/^{144}Nd$ values for acid-leached plagioclase from FT BH and FT 820 (table 12) are different from one another and from San Manuel porphyry. Because Nd isotopic values are thought to be rather insensitive to hydrothermal alteration (Farmer and DePaolo, 1987), the variation observed among initial $^{143}Nd/^{144}Nd$ values suggests that these are primary differences among the magmatic values, at least at the level of emplacement. The combined isotopic data suggest that Rice Peak porphyries have experienced a greater degree of (upper) crustal interaction than San Manuel porphyry, as also suggested by the zircon data.

PETROGENETIC CONCLUSIONS

The more varied character of the xenocrystic zircon in Rice Peak porphyry relative to San Manuel porphyry, as well as the trace element and isotopic signatures of these porphyries, are consistent with field relationships. At the sampled level of exposure San Manuel porphyry is enclosed within Oracle Granite that presumably extends to considerable depth and has a homogeneous population of zircon in terms of U-Pb age, whereas Rice Peak porphyries are enclosed within sedimentary rocks that may contain detrital zircon from a variety of sources.

The typical aphyric chilled margins of San Manuel porphyry (Force, this volume) may further suggest that long-lived crustal interaction was mostly at the level of emplacement. In contrast, Rice Peak porphyries seldom exhibit chilled margins, are choked with xenoliths, and show little evidence for thermal energy (Force, this volume). Magmas parental to Rice Peak porphyries probably ascended slowly from the site of melt generation, and may have interacted extensively with varied middle to upper crustal rocks. The data and conclusions presented here permit coeval or even cogenetic parental magmas for San Manuel and Rice Peak porphyries, but do not support detailed correlation.

Fluid Inclusion Studies of the Mammoth District

Robert J. Kamilli

INTRODUCTION

Force and Cox (1992) and Force (this volume, Appendix 1) suggested that the Mammoth Au-Ag-Mo-V-Pb-Zn-Cu deposit might be a detachment fault-related deposit instead of a polymetallic vein deposit as suggested by previous authors (Schwartz, 1953; Cox, 1986; and by implication Peterson, 1938, p. 44, 54). This section addresses this question, primarily by means of fluid inclusion studies. Specifically, it attempts to resolve the question, posed in Force and Cox, of whether the Mammoth deposit is an example of 1) intrusion-driven epithermal mineralization in a detachment-fault-system structural trap, or 2) involvement of a basinal brine that migrates upward along the detachment fault in a detachment-related thermal gradient. The latter alternative would be more typical of a true detachment-related deposit, as defined by Long (1992) and numerous other workers. Force and Cox also suggest the involvement of a second hydrothermal fluid, following the model of Spencer and Welty (1986).

Most samples for this study were collected by E.R. Force in the course of a study of the Mammoth and San Manuel districts (Force and Cox, 1992). Samples with the prefix NMNH were received from the Division of Petrology and Volcanology, Department of Mineral Sciences, Smithsonian Institution. All are listed in table 15.

PARAGENESIS

Two, somewhat different parageneses for the Mammoth vein system are by Peterson (1938) and Creasey (1950). This paper primarily uses Peterson's because it is better documented and more consistent with the findings of this study. Petersen's stages 1, 2, and 3 are clearly hypogene in origin and represent a continuum. Stage 1 is characterized by narrow quartz veinlets, generally cutting rhyolite and generally about 1 mm to 1 cm wide, with accessory adularia. Stage 2 is characterized by an abundance of specularite and chlorite with sphalerite and pyrite. Stage 3 is defined by the presence of galena, native gold, amethyst, and coarse comb quartz that was found in this study to contain primary fluid inclusions radiating from the C-axis toward the pyramid faces. Both Peterson (1938) and Creasey (1950) infer a distinct depositional break between stages 1–3 on one hand and 4 on the other.

Stage 4 represents the great mineralogical mystery (and rockhound's delight) of the Mammoth district. It is characterized by initial extreme leaching of earlier-formed minerals, leaving a porous quartz boxwork. This boxwork was subsequently filled by wulfenite ($PbMoO_4$) and vanadinite ($Pb_5(VO_4)_3Cl$), among many other minerals, including diaboleite ($Pb_2CuCl_2(OH)_4$, which will be discussed below. Although apparently supergene in macroscopic aspect and occurrence, they are paragenetically earlier than the indubitably supergene mineral assemblage that forms stage 5. More importantly, there is no evidence in the Mammoth vein system itself of previously formed Mo and V minerals, making it extremely unlikely that the wulfenite and vanadinite formed from typical supergene processes of oxidation or dissolution and reprecipitation.

Stage 5 is characterized by supergene minerals such as calcite, cerussite ($PbCO_3$), smithsonite ($ZnCO_3$), malachite ($Cu_2(CO_3)(OH)_2$), chrysocolla ($(Cu,Al)_2H_2Si_2O_5(OH)_4 \cdot nH_2O$), and hemimorphite ($Zn_4Si_2O_7(OH)_2 \cdot H_2O$). See Bideaux (1980) for an exhaustive list of Mammoth minerals.

FLUID INCLUSION STUDIES

Methods

Thin sections from 14 samples were examined petrographically for suitability for fluid-inclusion studies. Of these, 11 were chosen for heating/freezing and crushing studies. A total of 140 filling temperatures were measured and a total of 156 salinities were determined from these samples. All data are for quartz, except for diaboleite in stage 4. The measurements were made using a modified USGS gas-flow heating/freezing system manufactured by Fluid Inc. of Denver, Colorado, in conjunction with a Leitz SM-LUX-POL microscope[1]. Instrumental precision is ±0.1°C. No data were obtained for stage 5.

Three categories were used for classification of the fluid inclusions: Primary (P), Secondary (S), and Indeterminate (I). Primary fluid inclusions were identified on the basis of occurrence in crystallographically controlled growth zones, three-dimensional groups, or isolated occurrence. Secondary inclusions were identified on the basis of occurrence as planar groups. Indeterminate fluid inclusions are those that were thought to be primary on the basis of their similarity to other primary fluid inclusions in the same sample and the experience of the author, but for which a primary origin could not be conclusively demonstrated by petrography. All but six data points in figures 113 and 114 are from primary or indeterminate (i.e. probably primary) fluid inclusions.

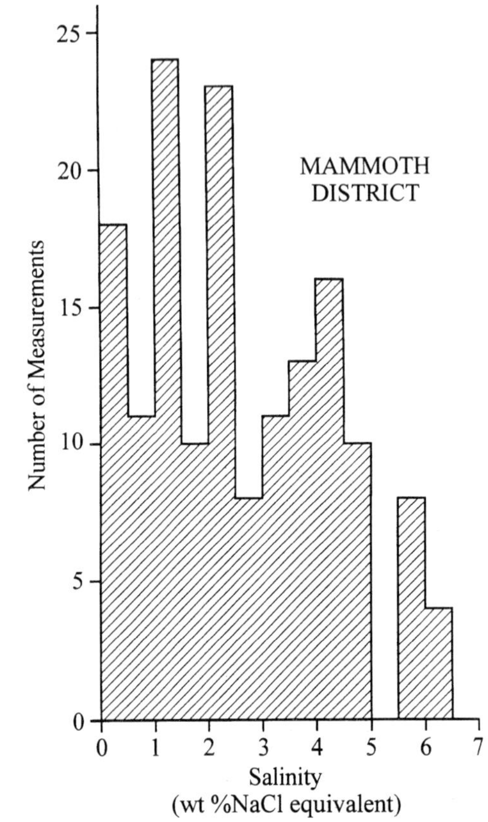

Figure 113 Histograms of fluid-inclusion salinities for all samples combined, Mammoth district.

General Observations and Crushing Studies

All fluid inclusions were liquid rich and two phase. No vapor-rich inclusions, CO_2-rich inclusions, or daughter minerals were observed. All samples were subjected to crushing in immersion oil while being observed under the microscope. None evolved any detectable CO_2 or other gases, such as CH_4. These observations were confirmed by the freezing studies, which yielded no evidence of clathrate formation.

Freezing Studies

Salinities of the fluid inclusions were obtained by measuring freezing-point depressions and are reported as weight percent NaCl equivalent. Measured salinities ranged from 0 to 6.1 weight percent NaCl equivalent for stages 1 to 3, and 0.2 to 2.1 for stage 4 (table 15 and fig. 115). Salinity histograms (fig. 113) suggest a bimodal distribution of salinities with modes at about 1.5 and 4.5 and a break at about 3 weight percent NaCl equivalent. This apparent bimodality is evident regardless of the inclusion types plotted (fig. 113). Nearly the complete salinity range was found in most individual samples (see especially samples MVTW, FT-864, and NMNH R10872 in table 15); one sample yielded only higher-range salinities (FT-146-B); three yielded only lower-range salinities (FT-748, NMNH R18086, and NMNH R18072-diaboleite).

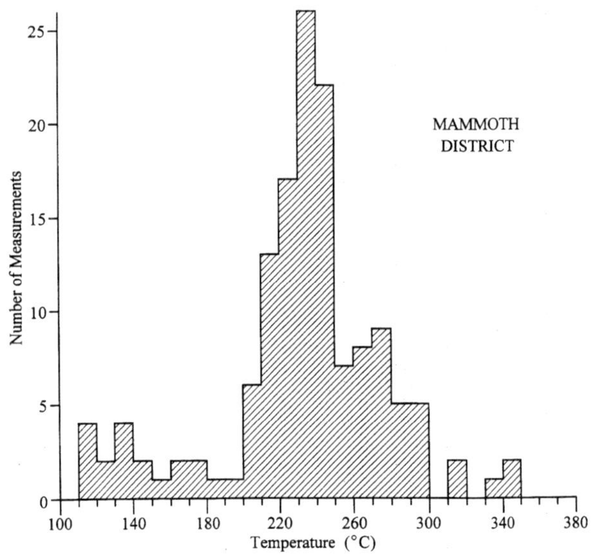

Figure 114 Histograms of fluid-inclusion filling temperatures for all samples combined, Mammoth district.

Filling Temperatures

The filling temperatures (table 15 and figs. 114 and 116) show a pattern typical of vein-type ore

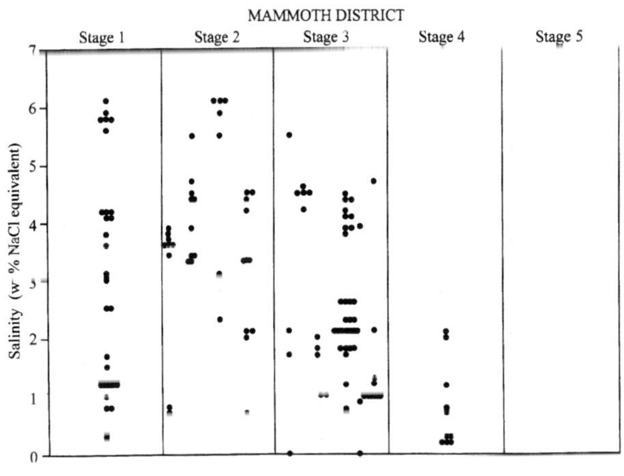

Figure 115 Fluid-inclusion salinities as a function of time (defined by paragenetic sequence), Mammoth district. Relative horizontal position within any stage is only a function of plotting convenience.

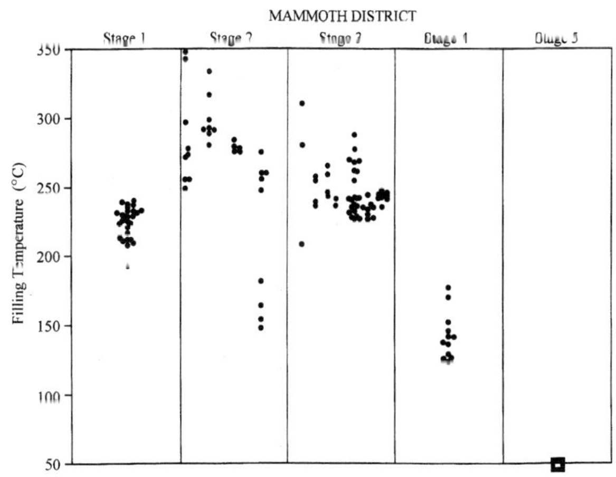

Figure 116 Fluid-inclusion filling temperatures of all measured inclusions as a function of paragenetic sequence, Mammoth district. Relative horizontal position within any stage is only a function of plotting convenience. Cerrusite indicated by open square.

Table 15 Fluid inclusion filling temperatures and salinities, Mammoth district

Sample	Vein	Stage (after Peterson, 1938)	Filling temperature °C (no. of measurements) [mean]	Salinity in weight percent NaCl equivalent (no. of measurements) [mean]
FT-864 (loc. 152)	Qz veinlets cutting rhyolite	1	193–240 (26) [224]	0.0–6.1 (29) [2.8]
FT-146-D (near loc. 56)	Collins	(1)*/2	249–347 (9) [285]	0.7–3.9 (9) [3.0]
"MVTW" (near loc. 132, stop 6C of Force & Dickinson, 1994)	Mammoth	2	275–284 (5) [278]	0.0–6.1 (8) [4.4]
FT-146-B (near loc. 56)	Collins	2	280–334 (8) [299]	3.3–5.5 (10) [4.1]
FT-146-C (near loc. 56)	Collins	2	147–275 (9) [260]†	0.7–4.5 (11) [3.1]
"Collins" (near loc. 57)	Collins	(2)*/3	228–289 (26) [250]	1.0–4.7 (28) [3.1]
FT-35 (loc. 133)	Mammoth	3	235–247 (10) [243]	1.0–4.7 (10) [1.5]
FT-748 (loc. 153)	Mammoth-related vein	3	227–242 (14) [235]	0.8–2.6 (13) [2.0]
NMNH R18086	"Mammoth-St. Anthony"	3	204–240 (17) [218]	0.3–2.6 (20) [1.1]
NMNH R18072 (Quartz)	"Mammoth-St. Anthony"	3	208–310 (3) [266]	0.0–5.5 (4) [2.3]
FT-146 (near loc. 56)	Collins	3/(4)*	245 (1)	0.0–3.9 (3)
NMNH R18072 (Diaboleite)	"Mammoth-St. Anthony"	4	124–177 (12) [142]	0.2–2.1 (11) [0.8]

* Parentheses indicates that sample had paragenetic evidence for the existence of that stage, but fluid-inclusion data were interpreted as being from the stage not in parentheses.

† Four anomalously low numbers disregarded.

deposits (Kelly and Turneaure, 1970; Kamilli and Ohmoto, 1977), i.e. an initial increase followed by a gradual decrease with time. However, in contrast to the pattern shown by Kelly and Turneaure (1970) and Kamilli and Ohmoto (1977), there was no evidence of boiling in the samples examined. The mean temperatures for stages 1 through 4 are 224°C, 283°C, 239°C, and 142°C, respectively. Four filling temperatures, ranging from 147–181°C in a single quartz grain in sample FT-146-C are anomalously low, and probably represent quartz formed during stage 4 in a sample comprised largely of stage 2 material. Bideaux (1980, p. 165) reports fluid-inclusion filling temperatures of about 50°C for stage 5 cerrusite.

Depth of Mineralization and Pressure Correction for Filling Temperatures

The lack of evidence for boiling in any of the fluid inclusions allows a minimum depth to be estimated for the Mammoth vein system using the data of Haas (1971). Assumptions in this calculation are: 1) the system NaCl-H_2O is a good approximation of the hydrothermal fluid; 2) the pressure gradient was hydrostatic; 3) no vapor bubbles were suspended in the overlying fluid column; 4) the temperature gradient was such that the overlying fluid column was everywhere at the boiling point. For 300°C (approximately the highest average filling temperature found), minimum depths of 980 and 1090 meters for 6 and 0 weight percent NaCl, respectively, are derived. This depth implies pressures of about 80–90 bars. Under these conditions, using the data of Zhang and Frantz (1987) and the FLINCOR program of Brown (1989), the calculated pressure is sensitive to small pressure corrections (the difference between the measured filling temperature and actual temperature at which an inclusion was trapped is called the pressure correction). Thus, a pressure correction only 10 degrees higher than the filling temperature approximately doubles the calculated pressure, and, under such circumstances, calculated depths rapidly become geologically unreasonable for this type of shallow deposit. A shallow depth of mineralization is also supported by association of these veins with apparently coeval rhyolite bodies that are both intrusive and extrusive in a system with structural and topographic relief of about 400 meters (Force and Cox, 1992). Therefore any pressure correction for actual temperatures of trapping is negligible compared to the scatter of filling temperatures found for each stage and therefore unnecessary.

DISCUSSION

Stage 4 Temperatures

Although filling temperatures for stage 4 diaboleite are relatively low at 124–177°C, these temperatures are clearly too high to be indicative of a supergene, weathering environment. This observation confirms Peterson's (1938) conclusion that minerals of stage 4, including wulfenite and vanadinite, had a hypogene origin. Wulfenite, in particular, was found in intimate association with diaboleite in the samples studied. It is clear, however, that much hypogene leaching of material from stages 1–3 took place before deposition of stage 4 minerals. Therefore, intimate association of different minerals is no guarantee that any given sample represents a single stage. This is demonstrated by the fact that the quartz in the diaboleite sample (NMNH R18072) yielded fluid-inclusion filling temperatures of 208–310°C and was deposited during stage 3.

Two Hydrothermal Fluids?

The bimodal distribution of fluid-inclusion salinities suggests that two hydrothermal fluids were present during deposition of the sulfide-specularite-chlorite assemblages characteristic of stages 1–3, one with a mean salinity of about 1.5 weight percent NaCl equivalent, the other with a mean salinity of 4.5 weight percent NaCl equivalent.

The two-fluid hypothesis is supported by a plot of salinity versus filling temperature (fig. 117). Points above the 3 weight percent NaCl equivalent line are within a relatively restricted range of 208–307°C, whereas points below that salinity value have a greater range of 124°C to 347°C. This way of expressing the data suggests that a break point between the two populations at 3 weight percent NaCl equivalent separates a low salinity fluid with a lower mean temperature and larger temperature variance from a higher salinity fluid with a higher mean temperature and a smaller temperature variance.

Origin of the Hydrothermal Fluids

It is impossible to definitively ascertain the origin of hydrothermal fluids without at least oxygen

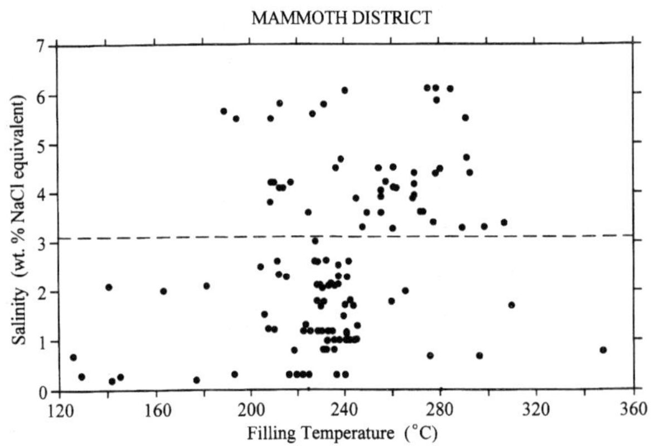

Figure 117 Plot of fluid-inclusion salinities as a function of filling temperatures, Mammoth district.

and hydrogen stable isotope data. Nevertheless, knowledge of the geology of a mineral deposit, along with fluid-inclusion data, allows for some reasonable interpretation. It is virtually certain that the low salinity fluid is a little-evolved meteoric water. Such is the case for all inclusion fluids from ore deposits formed in a nonmarine environment known to the author. See for example Kamilli and Ohmoto (1977), Casadevall and Ohmoto (1977), Bethke and Rye (1979), or Hayba and others (1985).

Likewise, most studies of similar, vein-type ore deposits have concluded that higher salinity fluids, such as seen in this study, are basinal brines (Roddy and others, 1988; Kerrich and Rehrig, 1987; Wilkins and others, 1986), also called "formation water" (Kamilli and Ohmoto, 1977), or "exchanged" or "evolved" meteoric waters (Bethke and Rye, 1979; Hayba and others, 1985). Studies of detachment-related ore deposits have shown that the basinal brines involved range from about 10 to 23 weight percent NaCl equivalent with later, distal veins having salinities of 6 to 12 weight percent NaCl equivalent (Wilkins et al, 1986; Spencer and others, 1988; Roddy and others, 1988; Allen, 1985; Long, 1992). Basinal brines do not, however, necessarily have salinities over 6 weight percent; see, for example, Hem (1970, p. 154, table 18) or U.S. Geological Survey and others (1969, p. 423, table 37). There are no known evaporites in the Cloudburst paleo-basin (Eric Force, oral communication, 1993) that could produce a high-salinity brine by interaction with circulating groundwater.

The following considerations are pertinent to an interpretation of the evolution of the hydrothermal fluids associated with the mineralization in the Mammoth district (from Force and Cox, 1992): 1) localization of mineralization, both with respect to zoning and the greatest quantities of ore, near flexures of the Cloudburst detachment fault (i.e. the intersection of the Cloudburst and Turtle faults); 2) occurrence of extensive epidote-chlorite alteration in the wallrock of the vein within 100 m of the Cloudburst detachment fault below; 3) vein formation coeval with cooling of rhyolite intruded during detachment faulting.

The following scenario is consistent with the data. Basinal brines of moderate salinity, derived from the recently formed Cloudburst basin, were heated by uplift of the lower plate of the Cloudburst detachment fault and migrated updip along the detachment toward the intersection with structures that are now the Mammoth vein set and ultimately into those structures. The lower plate acted as a relatively constant heat source, yielding a restricted filling-temperature range for the fluid inclusions formed from the basinal brines. Low salinity meteoric water also entered the Mammoth vein set periodically, influenced by episodic emplacement of rhyolite porphyry intrusions over an area of almost 1 km^2 that markedly varied the temperature of near-surface rocks over space and time, yielding fluid inclusions that varied widely in filling temperature. These two fluids were present during the deposition of minerals during stages 1 through 3. Mixing of the two fluids is not supported by the data reported here because of the lack of systematic changes in salinity and filling temperature (fig. 117). However, a line of mixing could be lost in the scatter of the data. Mixing of a reduced, metal-bearing basinal brine with oxygenated, near-surface meteoric water, as proposed by Spencer and Welty (1986), is an attractive mechanism for formation of the unusual stage 4 mineral suite at Mammoth, but does not seem to fit the data presented herein. This mineral suite may be a special case of the more typical oxidized assemblage associated with detachment-related ore deposits.

The apparent absence of brines during stage 4, when diaboleite, wulfenite, vanadinite, and manganese oxides were deposited, suggest that the meteoric fluids were indeed the oxidizing component of the hydrothermal system. More speculatively, the hydrothermal system associated with stage 4 may have increased with size over time, i.e. larger thermal anomalies and larger fluid fluxes, as has been shown for numerous deposits both in the field (Kamilli and Ohmoto, 1977) and theoretically (Norton and Cathles, 1979). This may be especially pertinent for Mammoth, where a series of individual rhyolite intrusions were emplaced over a period of time (Force and Cox, 1992). It is therefore possible that the hydrothermal system associated with the Mammoth district impinged on the San Manuel-Kalamazoo porphyry copper deposit about 1 kilometer to the south and scavenged molybdenum for stage 4 wulfenite from molybdenite and vanadium for stage 4 vanadinite from magnetite. This would explain the mystery, discussed by Peterson (1938) and Creasey (1950) of the source of molybdenum and vanadium for stage 4 ore at Mammoth. It is probable that the basinal brines involved in earlier stages never reached the San Manuel deposit because the Cloudburst detachment system, which apparently acted as the conduit for the brines, does not extend that far south. Molybdenite is found in the low-grade center of the San Manuel and Kalamazoo deposits, and both molybdenite and magnetite are found in the deep zones at San Manuel-Kalamazoo (Lowell and Guilbert, 1970). Magnetite is also present in the Middle Proterozoic Oracle Granite and in the San Manuel "A" porphyry (Lowell and Guilbert, 1970). The composition of magnetite in the San Manuel deposit is presently unknown.

Is the Mammoth Deposit a Detachment Fault-Related Deposit?

The Mammoth deposit has long been known to be highly unusual because of its exotic oxide mineralogy. It certainly is not a typical epithermal deposit or polymetallic vein deposit, as defined by Mosier and others (1986) or Cox (1986), respectively, although Cox classifies Mammoth as an example of the polymetallic vein type. Peterson (1938) suggested, and this paper confirms, that a major part of the oxide ore at Mammoth is hypogene and not supergene. The oxide minerals of stage 4 (Peterson, 1938) seem in the light of this paper to be simply a variant of the oxide mineralization so characteristic of detachment-related deposits (Roddy and others, 1988; Beane and others, 1985, 1986; Long, 1992). Indeed, the early sulfide assemblage and later oxide assemblage in the detachment related deposit model of Beane and others (1985, 1986) appears to be a close analog to the stage 1–3 mineralization and the stage 4 mineralization, respectively, at Mammoth.

The question remains as to whether Mammoth is actually related to the formation of detachment faults themselves. This was addressed by Force and Cox (1992), who did not reach a final conclusion, but made a strong circumstantial case for Mammoth being detachment related. A common theme in all previously published models of detachment-related deposits, is the upward migration of a basinal brine along the detachment and subsequent intimate involvement of this fluid in the formation of the resulting deposit. The fluid inclusion evidence presented in this paper suggests that this mechanism applies to Mammoth but that the basinal fluid is dilute. A complicating factor is the involvement of rhyolite intrusions and an associated second fluid, apparently a meteoric water. Epithermal deposits that simply occupy a detachment fault zone do exist (Welty and Spencer, 1986; Long, 1992), but these do not contain primary oxide minerals and elevated base metals. Given the appealing two-fluid model of Spencer and Welty (1986), it is surprising that so many detachment-related deposits contain apparently only one, saline (>10 weight percent NaCl equivalent) hydrothermal fluid. Perhaps hypersaline brines are characteristic of detachment-related deposits in southwest Arizona and not detachment-related deposits in general.

SUGGESTED FURTHER WORK

The results of this reconnaissance study need to be confirmed by further studies of fluid inclusions that are closely tied to a detailed study of the time-space relations of the mineralization at Mammoth. A strenuous effort should be made to measure fluid-inclusion filling temperatures and salinities of the semi-transparent minerals of stages 4 and 5, especially wulfenite and vanadinite. This might be achieved by use of an infrared microscope. Oxygen and hydrogen stable-isotope studies of minerals and inclusion fluids would test the two-fluid, meteoric-basinal brine hypothesis presented herein. Magnetite at the San Manuel-Kalamazoo deposit should be analyzed for vanadium.

CONCLUSIONS

The conclusions of this study are as follows:

1. At least two hydrothermal fluids were involved in the hypogene mineralization of the Mammoth-Collins vein system. One was low salinity (0–3 weight percent NaCl equivalent) and probably meteoric in origin; the other was higher salinity (3–6 weight percent NaCl equivalent) and probably a relatively low salinity basinal brine originating in the Cloudburst basin.

2. The wulfenite-vanadinite-diaboleite oxide assemblage (stage 4) that some authors have previously suggested as being supergene is definitely hypogene, with a mean depositional temperature for the assemblage of 142°C. Basinal brines were apparently absent during deposition of this assemblage.

3. Mean depositional temperatures for the quartz - chlorite - hematite - base-metal-sulfide assemblages, i.e. stages 1, 2, and 3, were 224°C, 283°C, and 239°C, respectively.

4. Hydrothermal fluids were essentially NaCl brines, with little or no CO_2, CH_4 or other organic fluids present. Boiling of the hydrothermal fluids did not occur.

5. The Mammoth deposit is best thought of as a variant of the detachment-fault related deposit type, with a characteristic hypogene oxide-mineral assemblage, close spatial and temporal association with a detachment-fault system, and the involvement of a relatively low-salinity basinal brine.

Acknowledgements

This paper benefitted from reviews by Keith Long and Eric Force.

Endnote

1. Any use of trade, product, or firm names is for descriptive purposes only,and does not imply endorsement by the U.S. Government.

Appendices

Appendix 1
Description of Mineral Deposits

Some sources I used for mine descriptions are unpublished but publicly available. These include 1) the Mineral Resource Data System (MRDS) of the U. S. Geological Survey, itself incorporating data from many sources, 2) a 1945 report by E. J. Ewing on the Campo Bonito district, available from the Buffalo Bill Historical Center, Cody, Wyoming, and 3) files of the Anaconda Collection, available from American Heritage Center, Laramie, Wyoming.

Mammoth District

Veins of the Mammoth set were mined discontinuously from 1879 to 1959, with about 2 million tons of ore removed. Effort was primarily directed at gold in the early years of production; the total recovery was about 400,000 ounces (Creasey, 1950). In later years, silver (about 1 million ounces), molybdenum (over 6 million pounds MoO_3), vanadium (over 2.5 million pounds V_2O_5), lead (over 70 million pounds), zinc (about 50 million pounds), and minor copper were the main products. In 1978, Magma Copper Company began mining the veins for silica flux, and recovered gold and silver.

General geology and past work. The Mammoth district was described by Peterson (1938), with later additions by Creasey (1950). These works focus on vein mineralogy and paragenesis. Detailed mineralogy emphasizing mineral morphology was described by Bideaux (1980), and references therein. Structural evolution of the district was treated by Force and Cox (1992). Kamilli discusses the implications of fluid inclusions in the veins in a separate section of this volume.

Geometry and grade of the veins are not well described. Individual veins may be as wide as 2m (6 ft), but are most commonly a few centimeters wide, confined to zones locally over 10m (33 ft) in width. Grades of gold were in the range 1 to 15 parts per million. In sulfide-rich zones, lead and zinc could each be over 5 percent (Creasey, 1950). Production figures in Creasey's table 5 imply that MoO_3 and V_2O_5 grades were about 2,500 and 1,000 ppm, respectively, in some segments of the veins.

The veins are emplaced along old faults (called vein-faults in Peterson, 1938) and consist largely of multiple stages of both quartz and entrained fault gouge. Early veins also contain adularia and later veins, calcite. Gold values were highest in finely banded quartz of intermediate age. Specularite-rich zones are characteristic, as are open-space-filling amethyst, wulfenite, and vanadinite. Manganese oxide, barite, and fluorite are common both in the veins and in joints in wallrock. A vertical zonation in the reconstructed vein set is shown in figure 102. Mineral paragenesis (Peterson, 1938 and Creasey, 1950) is further discussed by Kamilli (this volume).

The morphology of the veins of the Mammoth set is sufficiently distinctive that genetically related veinlets can be recognized outside the main vein set. Typically, the black specularite zone is present near both vein walls, separated from it by a thin white zone in which adularia can be recognized locally (fig. 104). The core consists of white (or locally purple) quartz.

The veins are approximately co-extensive with a zone of rhyolite intrusive and extrusive rocks that extends northwestward from Tiger area (loc. 133) across Tucson Wash (pl. 1). Some of the higher gold values were obtained where the veins cut these rhyolite bodies.

The Mammoth vein set now consists of two main segments, the Collins and the Mammoth-Tiger segments, on opposite walls of the Mammoth fault zone (pl. 1; Force and Cox, 1992, fig. 2). The veins dip steeply, mainly southwest, whereas the Mammoth fault dips steeply northeast (fig. 101). The Mammoth-Tiger vein segment to the east has apparently been dropped down on the Mammoth fault relative to the Collins vein segment to the west. The amount of separation was listed as about 700 feet (210m) by Creasey (1950). Force and Cox redetermined this figure as 264 $\pm$ 2 m (866 $\pm$ 6 ft).

In both segments, mining of the veins was mostly south of the Turtle fault, where the Oracle Granite and Cloudburst Formation are the main host rocks. The Collins segment was traced with little offset across the Turtle fault (locs. 56, 132) and northwestward for over a kilometer by Creasey (1965).

The Collins segment. The Collins vein segment is that to the west of the Mammoth fault zone, on the lower plate of the San Manuel fault system; that is, it is part of the same crustal block as the San Manuel porphyry copper deposit (pl. 1). The Collins vein segment trends northwest from its surface intersection with the Mammoth fault, through the Collins workings (loc. 57), and across the Turtle fault (loc. 132) with little offset. The vein shows about 160m (530 ft) of vertical relief on the surface. The host rocks are Oracle Granite and Cloudburst Formation south of the Turtle fault, and entirely Cloudburst Formation north of it. Minor mineralization is found on the Turtle fault itself. Workings described in the older literature are all south of the Turtle fault, but significant effort has clearly been expended on working the northern portion as well (loc. 56).

The Collins segment of the Mammoth vein set strikes about N. 40° to 50° W., and dips about 70° to the southwest; variations in attitude are shown in detail by Peterson (1938). Mineralization is stronger where strike is more northerly. The subsidiary Collins East vein (not shown) dips steeply northeast, and the intersection with the main Collins segment plunges gently southeast.

New exposures of the Collins vein segment are a result of open-pit silica mining in the area of the old Collins mine (loc. 57), on a hilltop at an elevation of about 3,300 to 3,400 feet (about 1020 m). The new exposures clarify relations shown by Peterson (1938) in this area. His mapping predated open-pit mining and relied on underground and natural-outcrop information.

The new exposures of the Collins vein show a gently east-dipping contact between underlying Oracle Granite and overlying conglomerate of the Cloudburst Formation on the northeast wall of the pit (Force and Cox, 1992, fig. 2). Both units are transected by younger rhyolite intrusions which must be mentally extracted to perceive the original relations. The elevation of the Oracle-Cloudburst contact at the vein is about 3,360 ft (1025 m); it is apparently an unfaulted unconformity. The southwest wall consists of Cloudburst Formation and rhyolite entirely, except at the south end where a large block of Oracle Granite is juxtaposed along a steep N. 70° W. structure. The fault into which the vein was emplaced was apparently east-side-up, with an interfering structure on the west side. Force and Cox show that dip-slip movement was about 35 m (120 ft). A right-lateral component of fault movement is also indicated by the distribution of open-space mineralization.

Where the Collins vein segment is in Cloudburst Formation host-rocks, both are saturated with epidote and chlorite, especially in fractures and vesicles. Associated veinlets are quartz and minor adularia rather than calcite. A marginal zone shows manganese oxides and (or) barite on joint surfaces. This style of alteration extends only about 100 m (300 ft) laterally from the Collins vein segment and associated rhyolite, and shows a ceiling at an elevation of about 3300 ft (1000 m).

The Mammoth-Tiger segment. The Mammoth-Tiger segment of the Mammoth vein set is east of the Mammoth fault zone. At depth this vein segment is truncated by the Mammoth fault zone.

In many ways the surface traces of the Mammoth-Tiger and Collins vein segments are similar. The Mammoth-Tiger vein dips southwest about 70°; mineralization is apparently strongest in the most nearly north-south-striking parts of the vein. At depth the Mammoth-Tiger vein intersects the subsidiary Dream vein, dipping northeast 45 to 65°; the intersection plunges gently southeast (Peterson, 1938). The veins are closely associated with rhyolite, mostly intrusive near the veins but locally extrusive in the same structural block. The Mammoth-Tiger vein segment intersects the Mammoth fault zone south of the Turtle fault, but its host rocks include both Oracle Granite and Cloudburst Formation. Alteration is chloritic but not epidotic in the Cloudburst host rocks of the Mammoth-Tiger vein segment.

As in the Collins vein segment, recent open-pit mining of the Mammoth-Tiger vein segment (loc. 133) has clarified relations shown by Peterson (1938). West of the vein-fault, only Cloudburst Formation is exposed, whereas on the east side, Oracle Granite is overlain by Cloudburst Formation conglomerate along a gently east-dipping contact (Force and Cox, 1992, fig. 2) whose elevation at the vein is about 3,200 feet (975 m). Both pit walls are cut by intrusive rhyolite. Peterson (1938) describes the contact in his text and on sections of the subsurface Mohawk mine workings to the south. Thus the Mammoth-Tiger open cut is similar to the Collins open cut in that the vein-fault is down on the west, and an Oracle-Cloudburst contact is exposed on the east wall of the pit. Exposure of this valuable relation in both cuts is apparently fortuitous. As in the Collins vein segment, the relation of mineralization intensity to strike of vein-faults suggests a component of right-lateral motion.

Reconstruction on Mammoth fault. Force and Cox (1992) calculated the offset of the segments of the Mammoth vein set as 262 to 265 m (860 to 875 ft) in purely dip-slip sense, east side down. They based their calculation on offset of the Mammoth vein set, coupled with offset of the Turtle fault, and apparent offset of the Dream and Collins East veins.

Fault movement dropped the Mammoth-Tiger vein segment from a former position atop the Collins segment. Reconstruction as shown in figure 102 implies that the two segments of the Mammoth vein set formed a once-continuous vein set about 550 m (1,800 ft) deep. An additional 640 m (2,100 ft) now eroded would have separated the portion still in the ground from the San Manuel fault system. The total strike length of the reconstructed vein set is at least 2.5 km.

The zonation in mineral assemblages of the reconstructed vein shown in figure 102 is derived mostly from descriptions of the separate segments by Peterson (1938) and Creasey (1950). The economic mineral assemblage of the veins is oxide-dominated except for the deepest 60 m (200 ft), where lead and zinc show their maximum concentrations as sulfides. At lower levels than those described by Creasey, underground mapping showed a return to oxide assemblages (Keith Long, oral commun., 1993; not shown in fig. 102). At shallower depths, lesser lead and zinc are present as carbonates. Molybdenum and vanadium values reach maxima 60 to 400 m (200 to 1,300 ft) above the deepest levels of mineralization as the oxides wulfenite (*not* molybdenite as stated on p. C27 of Force and Cox, 1992) and vanadinite, respectively. Probably tungsten also reaches a maximum here, as the wulfenite concentrates averaged 2 percent WO_3. Wulfenite and vanadinite are virtually absent at deeper levels. Copper, as carbonate and silicate minerals, shows a maximum at 180 m (600 ft) above the lowest level, but lesser copper sulfide increases downward. Gold distribution is irregular, and may be controlled by post-faulting redistribution. Peterson noted that the zonation outlined above is present along the strike of the veins also, such that the solutions seem to be derived from deep under the northwest end of the Collins vein segment.

Peterson (1938) and Creasey (1950) differed on the part played by supergene oxidation in the mineral assemblage of the veins, particularly wulfenite and vanadinite. Peterson regarded these minerals to be original though somewhat late-stage hydrothermal minerals whose concentration maxima represent a depth zone in the originally continuous vein, whereas Creasey believed the minerals are a result of weathering, partly before and partly after offset on the Mammoth fault. The reconstructed mineral zonation (fig. 102) makes sense without pre-fault weathering. Kamilli (this volume) finds that no supergene assemblages are present in fresh samples.

Relation to rhyolite. The Mammoth vein set is closely associated with a swarm of rhyolite bodies on both sides of the Turtle fault (pl. 1). The rhyolite is of course intrusive where Oracle Granite is the host rock, but in Cloudburst host rocks it appears to be both intrusive and extrusive. Most of the clearly extrusive rhyolite bodies have the characteristic stratigraphy and structure of ash-flow tuff, and are concordant to bedding in conglomerate of the Cloudburst Formation (loc. 138). Many of the intrusive (but not extrusive) rhyolite bodies are recrystallized to muscovite, quartz, and epidote at the expense of feldspar (loc. 94). This alteration was apparently deuteric, as the extrusives in the same area are not altered in this way. Near the Collins vein segment, intrusive rhyolite contains quartz veinlets that form 10 percent or more of the rock as boxworks (loc. 133). Each veinlet shows a simple version of the same zoning morphology shown by the Mammoth vein set (fig. 104). Quartz veining in other host rocks of the main vein does not reach this density. Toward the main Mammoth vein set, the veinlets in rhyolite merge and become more complex in zonation. It is therefore likely that cooling rhyolite was the source of silica in the Mammoth vein set; silica was probably released by the reaction:

$$\text{potassium feldspar} + H^+ = \text{muscovite} + \text{quartz} + K^+$$

for which there is petrographic evidence.

Relation to Cloudburst-Turtle detachment system. The Mammoth vein set is spatially closely associated with the Cloudburst-Turtle detachment system (pl. 1), and related contact-metamorphism (Force and Cox, 1992). Indeed minor Mammoth-style mineralization is found on the Turtle fault (loc. 132). These relations lead to the question of possible genetic relations between the two sets of features.

The Collins vein segment crops out near a window through the Cloudburst detachment (loc. 94), and the projected subhorizontal detachment intersects the part of the Collins vein segment north of the Turtle fault (figs. 95, 101) at an elevation of about 2,900 to 3,000 ft (about 900m). This underground portion of the Collins vein is undescribed. The intersection is just above the sulfide zone of the vein set (fig. 102). That is, the observed zonation in vein mineralogy may be centered on this intersection.

In the area north of the Turtle fault, Cloudburst Formation rocks that host the Collins vein segment are characterized by epidote-chlorite alteration or metamorphism, producing a strong induration of these rocks (loc. 56). The alteration south of the Turtle fault is difficult to distinguish from older Laramide propylitic alteration associated with the San Manuel deposit. Two styles of alteration were observed in the northern area: (1) Widespread but partial epidote-calcite replacement of conglomerate matrix and filling of volcanic vesicles. The upper limit of this alteration is a gently dipping plane. (2) Local but more complete alteration in the immediate vicinity of the Collins vein segment. This alteration although vertically oriented shares its upper boundary with alteration of the first type. The first style of alteration probably reflects high local geothermal gradients and fluid movement induced by tectonic denudation on the Cloudburst detachment fault where it is present at depths of 100 m (300 ft) or less (fig. 95). The second probably represents heat and fluid transport laterally from the vein system, which acted as a conduit, but in the thermal environment controlled by detachment. Cooling rhyolite and other shallow intrusions were probably additional sources of heat for alteration.

Emplacement of the Mammoth vein set closely followed Cloudburst detachment faulting, accumulation of the Cloudburst Formation volcanic rocks and conglomerate, and rhyolite emplacement. That is, the vein set cuts the Turtle fault, the Cloudburst Formation, and rhyolite bodies. However, its silica apparently came from cooling rhyolite, and the vein-associated alteration aureole was a response to heat and fluids from cooling rhyolite and detachment faulting. We know independently from geologic relations that rhyolite emplacement was synchronous with detachment faulting. Mineralization was complete by the time of initial deposition of the San Manuel Formation at 22 Ma, only about a million years after rhyolite intrusion.

Therefore the relation among mineralization, detachment faulting, and rhyolite emplacement is not only spatial but temporal and physical. A genetic relation seems inescapable. This relation is broadly similar to that described by Spencer and Welty (1986) and Roddy and others (1988) for detachment-related deposits. In such deposits, mineralization forms along detachment surfaces and related steep normal faults by circulation of hot basin-derived fluids during fault movement, in response to tectonic denudation. Force and Cox (1992) listed the similarities and dissimilarities of the Mammoth vein set with detachment-related deposits of the southwestern U. S., in terms of mineralization and relation to wallrock features, and concluded that two possible forms of the relation are 1) rhyolite-driven mineralization in the chemical and physical environment conditioned by detachment, and 2) precipitation at the boundary of two fluids, one supplied by rhyolite and the other transported along the detachment fault. Kamilli in this volume addresses this question using fluid inclusion studies. Possibly the combination of factors at Mammoth can be considered a valuable intersection of mechanisms that produce metal enrichment. Implications for mineral resource potential are brought out in the section of that name.

Outlying deposits. The Mammoth district is here considered to include several small deposits in the same area that seem related in terms of origin and age.

Tar (or Tarr) mine is in a wash of the same name (pl. 1). Wilson (1941) described these workings soon after they were initiated, and Creasey (1965) described them after mining had ceased (probably in 1943). Additional description was gleaned from MRDS record M050190 (table 1). Shallow workings were in flat faults containing scheelite. At some point a quartz vein striking N. 70° E., dipping 80° southeast was worked for scheelite, wolframite, and copper minerals. Creasey (1965) showed an associated fault striking N. 35° W, dipping 60° northeast. Eventually the shaft became "several hundred feet" deep, probably with lateral workings, and passed from volcanic rocks of the Cloudburst Formation into Precambrian granodiorite (mode given by Creasey, 1965, table 2). Figure 95 suggests that this contact is at the Cloudburst detachment fault, at a depth of no more than 15 m (50 ft), and this shallow depth figure is supported by the high proportion of granodiorite in the dump. Perhaps as much as half of this granodiorite is a chloritic breccia. Thus the typical geology of the Cloudburst detachment fault is intimately related to the Tar mine, suggesting that the deposit is related to detachment and thus to the Mammoth district. Reported production was only 25 pounds of tungsten concentrate (Dale, 1959).

Pearl (Cuba) mine is outside the study area to the north about 1.5 km. It produced 50,000 to 60,000 (short?) tons (Schwartz, 1953) of ore containing copper oxide and sulfide, lead, silver, and gold from 1915 to 1941, from quartz veins along shear zones in granodiorite. These veins strike about N. 20 to 35° W., and dip steeply southwest. This is the attitude of veins of the Mammoth set, and since the morphology of the veins is also similar to the Mammoth veins, they are likely to be related.

Ford mine (pl. 1) in Tucson Wash is along a fault and quartz veins with typical Mammoth-vein attitudes. Oracle Granite and diabase (?) wallrock show epidote and manganese oxide alteration; the character of mineralization is not known. Force and Cox (1992) showed that the Ford mine on the lower plate of the San Manuel fault probably correlates with the Shultz Spring altered area in Cloudburst Formation conglomerates on the upper plate.

No discrete veins are present in the Shultz Spring area (pl. 1), which shows diffuse concentric alteration zones elongate NNW-SSE, ranging from pyrite at the center through specular hematite to chlorite. Barite and manganese oxides are ubiquitous, and adularia is locally present. Surface geochemistry in the Shultz Spring area showed a weak but well-centered anomaly of barium, silver, molybdenum, lead, gold, and copper, with an associated offset anomaly of zinc (Force and Cox, 1992). Altered Cloudburst Formation is unconformably overlain by San Manuel Formation. Taken together, Ford mine and Shultz Spring alteration suggest mid-Tertiary mineralization rooted in a plane with an attitude typical of the

Mammoth vein set. Hence they are included in the Mammoth district.

Creasey (1965) mapped and described ("sections 20, 21, and 28", p. 58) weakly mineralized faults in Cloudburst Formation and granodiorite extending intermittently about 2 km. These structures post-date detachment and strike northwest, dipping steeply in either direction. Creasey also described ("central part of section 17", p. 59) weak mineralization in a fault that appears to be a tear in the Cloudburst detachment. These all seem likely to be related in age and origin to the Mammoth veins.

Hatch (Harvey Hatcher) mine in Cloudburst Formation about 2 km south of Signal Peak is said (MRDS record M899859) to show copper and manganese oxides in north-south calcareous veins associated with a small porphyritic intrusion. They were worked in an open cut measuring about 10 x 15 m, 10 m deep. The deposit is apparently about the same age as Mammoth veins, but otherwise the relation is tenuous.

Prospects Related to San Manuel Fault

At many points along the San Manuel fault, prospect pits reveal weak manganese- and copper-oxide mineralization. Creasey (1965, p. 59) described two of these under "other prospects" (2nd and 3rd paragraphs). Two others form MRDS records M050183-4, near Signal Peak. A similar occurrence (MRDS record M050187) is along the related Red Rock fault near Mammoth Wash. Where the San Manuel fault crosses Cloudburst Wash, the area between splays contains fractures with veins of calcite and fine silica containing chrysocolla. These occurrences probably reflect solutions moving along the San Manuel and related faults from copper-rich bodies such as San Manuel ore below, and are not thought to be of importance.

San Manuel (+Kalamazoo) District

The San Manuel porphyry copper deposit was discovered in 1943-1944 by drilling near a surface outcrop at Red Hill. It has been a major producer since 1955. In 1989, the most recent year for which San Manuel was the sole contributor, Magma Copper Co. reported production of 377 million pounds of copper from a combination of underground block-caving operations, an open pit, and an in-situ leaching operation, all in the same orebody. Grade has averaged about 0.8 percent copper.

The Kalamazoo porphyry copper deposit, discovered in 1965-1966 (Lowell, 1968), is an offset segment of the San Manuel deposit. The orebody is entirely concealed, and its discovery by deep drilling resulted from geologic reasoning concerning the geometry of the San Manuel orebody and associated alteration zones, and movement on the San Manuel fault. Developmental production of the Kalamazoo deposit by block-caving began in 1990 by Magma Copper Co.

General geology. The San Manuel and Kalamazoo segments of the original deposit have been separated by the San Manuel fault, with the Kalamazoo segment down the fault dip of 25 to 30° toward the southwest. Lowell (1968) gave the figure for separation as 8000 ft (2440 m) in a direction of S. 55° W., modified by Force and Cox (1992) to 2363 m (7750 ft) toward about S. 46° W.

The basic geology of the San Manuel segment was described by Steele and Rubly (1947), Schwartz (1953), Pelletier and Creasey (1965), and Thomas (1966). The Kalamazoo segment has been described by Lowell (1968). Alteration of both segments was described by Pelletier and Creasey (1965) and Lowell and Guilbert (1970). Force and Dickinson (1994) and Force and others (1995) describe the geometry of intrusion and its tectonic restoration.

The pre-fault geologic relations of the San Manuel and Kalamazoo segments are the same. Oracle Granite, with lesser aplite and diabasic dikes, formed the host rock to multiple intrusions of San Manuel porphyry in the Laramide. The intrusion that seems to be associated with ore and alteration (porphyry A of Lowell and Guilbert, 1970) is tabular (Force and others, 1995), with aspect ratios of about >3 (length) : >3 (height) : 1 (thickness). That is, it is known to be present for a horizontal distance of 900 m (3000 ft), down-dip for about 900 m (3000 ft), and is approximately 300 m (1000 ft) thick. The present dip of this body is about 47° to the southeast, about a strike of about N. 58° E. in the longest unbroken segment from the San Manuel fault to the West fault (Force and others, 1995). Segments to the east have slightly different attitudes, and that of the Kalamazoo deposit has not been published. The tabular productive intrusion at present has a sub-horizontal bottom, at an elevation of about 1200 ft (365 m). Other, presumably pre-ore, intrusions of San Manuel porphyry, such as porphyry B of Lowell and Guilbert (1970), overlie this tabular intrusion. Some of these, like the one exposed in the Purcell window (loc. 72), have entirely different trends.

Toward the north and east of both segments, the deposit and its host rocks are overlain by Cloudburst Formation (loc. 26). Elsewhere, they are overlain by San Manuel Formation (pl. 1). San Manuel Formation also structurally overlies ore of the San Manuel segment along the San Manuel fault (fig. 98). Large portions of ore in both segments are under outcrops of Tertiary formations. Dikes correlated with Cloudburst Formation volcanics cut ore, but may be permeated with copper oxides.

Ore. Ore forms an envelope around certain parts of the productive porphyry intrusion. Two main limbs are present, 300 to 500 m (1000 to 1600 ft) apart (Schwartz, 1953; Pelletier and Creasey, 1965); both dip southeast with the same attitude as the igneous body. Indeed, the envelope itself constrains location of the *upper* surface of the tabular productive porphyry (porphyry A of Lowell and Guilbert, 1970) within undifferentiated San Manuel porphyry. The envelope also has an upper and a lower cylindrical termination (fig. 99) joining the two limbs. The lower termination plunges about 30° southwest, i.e. it departs from the lower termination of the tabular igneous body at an angle of about 30°. The upper termination plunges gently northeast; a large share of this termination is in the Kalamazoo segment. The tabular porphyry intrusion projects upward past this upper ore termination in the San Manuel portion. The upper and lower terminations converge toward the northeast but are truncated by the Cholla fault west of their projected intersection.

Sulfide ore averages 0.8 percent copper, with slightly higher values in Oracle Granite host (Thomas, 1966). The ore minerals in the sulfide zone are chalcopyrite and molybdenite, with minor bornite. Small amounts of silver and gold are present. Pyrite occurs in ore and is abundant in an outer zone. Oxide ore overlies the sulfide zone and is thicker toward the east. It contains chrysocolla, chalcocite, and lesser covellite, cuprite, bornite, and native copper. Data in Schwartz (1953) suggest that the base of the oxide zone plunges eastward at about 25°. Supergene enrichment at the base of the oxide zone has been modest in this deposit.

Alteration. The spatial arrangement of alteration intensity at San Manuel-Kalamazoo was described by Pelletier and Creasey (1965) and Lowell and Guilbert (1970). These authors recognized a central potassic (K-feldspar-biotite) zone, flanked by phyllic (quartz-sericite) and propylitic (epidote-chlorite-carbonate) zones. Argillic (quartz-kaolinite-chlorite) alteration is minor, between phyllic and propylitic zones. At the western end of the Kalamazoo segment, the propylitic zone passes into magnetite-bearing rock, and the core of the potassic zone contains the assemblage K-feldspar-quartz-sericite-chlorite. Sulfide ore tends to follow the boundary of potassic and phyllic zones, and the outer pyritic zone is mostly in the phyllic alteration zone.

Structural reconstruction. Force and others (1995) show that restoration of the San Manuel and Kalamazoo deposits requires their rotation of 30 to 35° up to the northeast about an axis trending N. 40° W. in conjunction with reconstruction on the San Manuel faults. This produces a plane striking about N. 81° E., dipping 62° southeast for the tabular productive porphyry body, and probably for the plane of symmetry of the ore envelope as well. After reconstruction, the bottom envelope termination of the orebody plunges about 60° to the southwest, whereas the base of the tabular productive porphyry body plunges about 30° in that direction. The top envelope termination plunges gently but variably southwest also. The original top of the orebody matches the location of thickest oxide-zone ore, the present upper east corner. The K-feldspar-quartz-sericite-chlorite and magnetite-epidote-chlorite alteration zones both occur at the lower west end, where the ore envelope is wider and the emplacement depth greatest.

American Flag District

This district name I have given to deposits hosted by Oracle Granite, scattered over an area of about 8 square kilometers between Oracle and the Mogul fault. Older district names in this region (such as Old Hat and Oracle) have been misused to include larger areas.

In most of the deposits, mineralization is in quartz veins striking approximately north, along sheared contacts between Oracle Granite and east-dipping diabase sheets. Hematite and siderite gangue are locally present. The Ewing report (table 10) stated that the ores plunge northward at a moderate angle. Most of the deposits were worked for gold, but in some deposits the most valuable products were W-Mo-V or Cu-Ag-Pb. The age of mineralization is not known; I suspect that some is pre-Laramide, like similar deposits in northwestern Arizona.

American Flag mine. These deposits (loc. 139), worked intermittently from 1878 to 1961, are along a silicified sheared contact between Oracle Granite (to the west) and an overlying diabase sheet striking N. 15° W., dipping 32° east. A northeast-striking fault is shown there by Creasey (1967). Chalcopyrite and galena were separated, and copper, silver, lead, and minor gold were recovered. Production figures are not published, but total value of products is reported as $19,239. Workings include 80 m (265 ft) of inclines and drifts.

Bearcat (Lovejoy, Madre del Oro, Silver Cliff) mine. These deposits (loc. 140) are developed along the same set of diabase sheets (pl. 1) as at the American Flag mine, which are north about 300 m (1000 ft). The rocks are faulted here also along a northeast trend (Creasey, 1967), and Oracle Granite has been bleached by alteration. Two quartz veins at the Bearcat property strike N. 10° E., and dip 45 to 50° east. The eastern vein was the main producer of scheelite with lesser vanadinite and wulfenite. Minor gold and silver were present. Pyrite, calcite, and iron oxide accompanied ore. The workings include an open cut on the east vein 15 m (50 ft) long and 5 m (15 ft) deep, and 140 m (465 ft) of adits and inclines. Most of these workings have since been regraded. Production in 1939-1943 was 270 kg (600 lbs) of concentrate.

Highjinks (Carolina Moon) mine. Workings here (loc. 27) were a string of pits along a 10-15 m diabase sheet striking N. 20° W., dipping 40° northeast. Quartz veins along the contact with Oracle Granite contained gold and scheelite. Several thousand dollars in gold were produced.

Madrugador mine. Workings here (loc. 141) follow a mineralized fault striking about N. 40° W., dipping steeply northeast, with some quartz veining. The fault in Oracle Granite offsets a diabase sheet, apparently the southern extension of the large sheet at American Flag settlement. Workings seem to be localized where the fault cuts diabase and where younger east-west faults occur. Gold was recovered, and some galena was present. Mining extended intermittently from about 1940 to 1980.

North Cross workings. On the southern outskirts of Oracle (loc. 142) are prospects reportedly for copper and silver, in a north-dipping zone of silicification in Oracle Granite. Production is not reported and former workings have mostly been re-contoured.

Deposits south and west of American Flag Hill. Some workings here are in siderite-quartz veins along a steep northwest-striking fracture in and near the base of the thick diabase sheet forming American Flag Hill. The workings are undescribed and unnamed, but the Ewing report mentions gold and silver recovery in the area.

Deposits on Oracle Hill. A gently east-dipping diabase sheet forming Oracle Hill has a 1-meter quartz vein along its basal intrusive contact with Oracle Granite. This vein was extensively worked, but production is not known. According to local residents, these workings were once called the Oracle mines, and gave their name to the town and thus the granite body.

Other deposits. This district contains many small workings, mostly named; miners and geographers must have been fond of the area. All workings are in Oracle Granite, probably along diabase contacts. Among these mines and prospects are the following (with commodities and section numbers in T. 10 S., R. 16 E.): Santa Rosa (Au, Ag, Pb; 16); Juniper (Ag, Pb; 17); New Years and Golden Contact (Au, Ag, Pb; 9 and 16); and Halloween (Spook; Au, Ag; 7).

Campo Bonito District

I use the term Campo Bonito district to refer to deposits along and immediately south of the Mogul fault in the vicinity of the former mining camp of Campo Bonito (pl. 2C). Deposits north of the fault I have described as American Flag district, and those farther south as Southern Belle and other districts. The districts thus defined are spatially fairly distinct. From Laramide time onward, all three of these districts probably shared the same "plumbing" system, related to the proto-Mogul fault. My classification parallels established conventions in all respects but one, which is the inclusion of the *Old* Maudina mine in the Southern Belle district.

The Campo Bonito district was largely a tungsten producer, active intermittently from about 1913 to 1953, but most active during World War II. The most important deposits are in a small fault-bounded block of Escabrosa Limestone south of and adjacent to the Mogul fault (pl. 2C), in which high-grade replacement deposits of scheelite formed. This block contains the Morning Star deposit, and the Pure Gold workings of the Maudina mine holdings. These deposits were described by Bateman and Erickson (1944), Dale (1959), and Creasey (1967). My own observations have been incorporated in the following descriptions.

Morning Star mine. This deposit (at Bonito mine of pls. 1, 2C), operated intermittently from 1913 to 1953, is in a mass of silicified Escabrosa Limestone. Scheelite is concentrated along faults striking northeast and northwest, in zones up to 0.6 m (2 ft) wide containing as much as 10 percent WO_3 (MRDS record M050218). Disseminated scheelite is less than 1 percent in grade. Lead carbonate is present. At the portal of a lower opening is a partially silicified calcite-cemented breccia 1.5 m (5 ft) wide in limestone containing fine-grained scheelite. Workings include a glory hole 4 to 15 m (19 to 50 ft) deep measuring about 14 x 18 m (65 x 45 ft), and a total of about 140 m (450 ft) of adits, inclines, and shafts. Production has been estimated as over 6000 short tons of ore averaging one percent WO_3.

Pure Gold workings. Mined in 1943 and 1944, these workings form a part of Maudina mine holdings. Descriptions of the "Maudina" mine before 1943 refer to the Old Maudina mine but after 1943 refer to the Pure Gold workings, which adjoin the northern boundary of the Morning Star mine claim.

The large surface workings are in a mass of silicified limestone (apparently separate from that of Morning Star mine), locally brecciated and pervasively veined with quartz, that dips southwest 40 to 50°. These workings are 60 m (200 ft) long and 9 m (30 ft) wide. In addition there are 90 m (300 ft) of adits and inclines. Production from these workings were 6,284 tons of ore averaging about 1.5 percent WO_3.

Bateman and Erickson's (1944) cross-section A-A' shows granitic rock dipping southwestward under these workings at an angle of only about 40°. They regarded the contact as the Mogul fault, as did Creasey (1967) in his text but not on his plate. Both the dip and the strike are wrong for the Mogul fault, and I think that the structure shown by Batemen and Erickson is some combination of a Rice Peak porphyry dike as shown on plate 2C, and/or a gently dipping young fault that has offset the proto-Mogul fault at depth in the same manner that steep faults offset it in map view.

Cody tunnel. William ("Buffalo Bill") Cody owned this deposit (pl. 2C, loc. 122) and others in the district from 1909 to 1916. It apparently is called a tunnel in reference to lack of ore. The tunnel is 262 m (860 ft) long, with 100 m (335 ft) of crosscuts.

One meter north of the portal of the Cody tunnel, the proto-Mogul fault is exposed, striking N. 55° W, dipping 85° southwest. It is between protoclastic Oracle Granite to the north, and Dripping Spring Quartzite containing a diabase dike to the south. The tunnel, reported to be following a south-dipping vein, appears to intersect the proto-Mogul fault within a short distance. The Ewing manuscript reported 0.15% WO_3 and some gold in gouge and breccia.

Related deposits. Several other deposits that are outside the Campo Bonito district are found along the Mogul fault. The largest of these is at the western end of the exposed Mogul fault, about 4.5 km outside the study area, at the Little Hills mines. These have intermittently produced Au, Ag, Cu, Mo, Zn, and Pb from the 1880's to the present, from sheared rock associated with the fault.

The Dead Bull (Bluebird) mine, near American Flag Spring (pl. 1), produced Au, Ag, and Cu (as chalcopyrite) in 1937 and 1938 from quartz veins up to 1.3 m (4 ft) thick along diabase dikes in Oracle Granite (MRDS record M241184). Inclines and shafts total over 70 m (240 ft). Production is unknown. The reported location (pl. 1) suggests that the workings are in the Mogul fault zone adjacent to American Flag Formation; indeed, the Ewing report stated that the best ore in this and the adjacent Pagaoro deposit was in fault gouge.

Small unnamed prospects just north of the neo-Mogul fault along Oracle Ridge follow steep diabase dikes and sheets striking north-south and east-west. The diabase is severely altered to chlorite and foliated; fractures contain hematite and siderite. Locally the diabase has developed a granophyric facies and elsewhere contains granitic xenoliths. From the Oracle Ridge divide east in Bonito Canyon to a former mining camp (Taraldson claim?) are more workings in protoclastic Oracle Granite. Production information is lacking for this sub-district. Probably it is related to the neo-Mogul fault rather than the proto-Mogul fault as are the main deposits of the Campo Bonito district.

Southern Belle District

Deposits of the Southern Belle district here include the deposits south of the Mogul fault but probably influenced by its "plumbing system", in the area of the Southern Belle mines, including the Old Maudina mine (pl. 2C). The latter is included because its geology and location are more closely related to that of the Southern Belle district than that of the Campo Bonito district. Outside the study area but closely related in geology are the Burney mines of the Canada del Oro drainage. Also described here are placer deposits in the canyon below Southern Belle mine, here called Southern Belle canyon.

Southern Belle mines. These deposits (pls. 1, 2C), currently being explored by Newmont Mining Co., were intermittently worked for Au, Ag, and Pb from 1885 to 1964. Sub-horizontal to gently northeast-dipping quartz veins contained pyrite, galena or cerussite, and gold. They are developed in shear zones along and just below a sheared unconformity between the upper member of Dripping Spring Quartzite with some diabase below, and Campo Bonito Formation and Bolsa Quartzite above (pl. 2C). The Campo Bonito Formation is locally cut out. Some of the exploited horizons look like sheared paleosols (fig. 105). Some shafts are apparently collared in Bolsa Quartzite but work lower horizons. Northwest-striking steep cross-faults apparently post-date mineralization, and produce alternating horsts of Dripping Spring Quartzite and grabens of Bolsa Quartzite or Campo Bonito Formation on the ridge above.

Production prior to 1906 from a single 50 x 150 m (160 x 500 ft) stope along a 2 m (6 ft) quartz vein is reported as about 18,000 tons (Heylman, 1989). Other (remaining?) vein material is reported to be over 30 times greater. The Ewing manuscript recorded minor scheelite in adjacent workings.

Placers in Southern Belle canyon. Detrital gold, presumably derived from the Southern Belle deposits, has been mined in placer deposits of Southern Belle canyon intermittently from about 1884 to the present. These are among the most productive placer deposits of the study area. Johnson (1972) reported that some of the richest workings were on a weathered diabase surface. An enrichment in detrital gold farther downstream is apparently related to the Mogul fault.

Old Maudina mine. This deposit (pls. 1,2C) was worked during 1915-1916 and 1943-1944, mostly for tungsten as scheelite. The cleanest descriptions are by Wilson (1941) and Creasey (1967); in descriptions made before 1943, the Old Maudina is referred to as the Maudina, whereas in most post-1943 descriptions, the term Maudina refers to the Pure Gold workings of the Campo Bonito district.

The deposit is along a fault that strikes approximately east-west, dipping 80° to the south at the mine, swinging northwesterly toward the west (pl. 2C). It offsets the Southern Belle Member of the Abrigo Formation in a complex manner. The ore shoot is described as plunging 45° east below a quartzite layer (Wilson, 1941), apparently in the fault and along its intersection with the northeast-dipping base of the Southern Belle Member. Ore ranges from 1.2 to 4.5 m (4 to 15 ft) in width down-plunge, and is a quartz-scheelite vein, with wulfenite and cerussite in upper portions. In some lower depth intervals, scheelite is practically free of gangue. Workings include a long stope along the plunge, a shaft 53 m (175 ft) deep, and about 300 m (1000 ft) of crosscuts and drifts. Production implied by Dale (1959) and Creasey (1967) is about 46 tonnes (102,000 lbs) of WO_3.

Related deposits. The Burney mining district of the Canada del Oro drainage just west of the study area is developed along contacts of Dripping Spring Quartzite with overlying Mescal Limestone, mostly in the southwestern limb of the Oro syncline (fig. 3). Veins along steep faults parallel to the Mogul fault are an important ore control. The deposits produced Pb, Zn, Au, Cu, and some barite and manganese oxide, from 1947 to 1967.

In Canada del Oro are many former gold placer operations in gravel, mined mostly in 1931 and 1932. Some of the placers were probably derived from primary mineralization of the Burney district.

Mary West (3C Ranch) Mine

This deposit (loc. 118), worked 1955-1960 for Pb, Cu, Zn, Au, and Ag, is along the intersection of the contact of the middle (sandy) and upper (pelitic) members of the Pioneer Formation with converging steep faults. One striking N. 20° E. is the locus of many prospects, both north and south of Mary West mine. Mineralization consists of quartz veins with galena, pyrite, siderite, and epidote. Morphologically similar veins cut intrusive Rice Peak porphyry (older facies), locally cleaved, to the south.

The veins, up to 1.3 m (4 ft) thick, were worked along a 95 m (310 ft) incline plunging east and an associated stope. Production figures are not known. Files of the Anaconda collection suggest that some production under both the Mary West and 3C names was from separate workings along the Mogul fault.

The map pattern (pl. 1) suggests that the N. 20° E. fault is pre- or syn-Rice Peak intrusion, and the distribution of veining also suggests a relation to Rice Peak. No other deposit in the study area has as plausible a link to the older facies of Rice Peak porphyry—the Campo Bonito district contains a single dike probably of the younger variety of Rice Peak porphyry, but with no evident relation to mineralization; the Leatherwood mine group of the Marble Peak district is associated with a fault intruded by the younger variety of Rice Peak porphyry.

Lower Nugget Canyon Prospect

In Nugget Canyon below the Control road, in the upper member of American Flag Formation, is an area of alteration located in the course of this study (loc. 145). Propylitic alteration is overprinted by local bleached sericite-pyrite alteration. Quartz-chlorite veins strike northeast at a high angle to bedding.

Alteration clearly post-dates some Rice Peak porphyry, as cobbles of porphyry occur in altered conglomerate. L. J. Cox and T. Harms (U. S. Geological Survey, oral commun., 1993) report that a prospect based on six claims has been established in the area.

Other Deposits between Mogul and Geesaman Faults

Other prospects in this area are few and poorly described. They include the following (with host formations and section numbers in T10S, R16E):

Poor Man (Abrigo; 29)
Circle J (Abrigo; 33)
Little Hill claim (Campo Bonito-Dripping Spring; 23 in R15E)

Silica for flux was apparently mined in section 30 (T10S, R16E).

Deposits along the Geesaman Fault

A string of deposits and occurrences extending about 8 km along the Geesaman fault (pl. 1) has not previously been grouped as a district; some had been included with the Marble Peak district based on proximity but not geology. However, the Geesaman fault appears to me to be an essential factor in the formation of this string of deposits (and it is not for the Marble Peak deposits). These deposits form a zoned population grading from previously described greisens to the west, through tungsten deposits, to base- and precious-metal deposits on the east, and to be related to Catalina granite in age and origin. This population could be named after its essential feature, the Geesaman fault, but doing so would create confusion with the Geesaman mine of the Marble Peak district.

Taylor X (Catalina Camp) workings. These undescribed adits and shafts (pl. 1) are developed along quartz veins in Abrigo Formation that forms a large inclusion in Leatherwood granodiorite, about 50 m south of the Geesaman fault. The Abrigo is altered to garnet-epidote assemblages, and then mylonitized; mylonitic foliation and veins strike N. 80° E. and dip 70° south. The pre-Leatherwood Geesaman fault shows some post-Leatherwood reactivation at this locality; Leatherwood is mylonitized. Two other possibly pre-Leatherwood faults (pl. 1) are also nearby and could have contributed to the "plumbing system" for this deposit.

Exploitation was for tungsten and copper as scheelite and chalcopyrite. Production in the early 1900's is unknown but from about 1948 to 1953 was about 50 tons of ore, some hand-sorted. No road to the deposit was constructed but at present a 4-wheel-drive road passes near it.

Corregedor mine. A shaft in pyroclastic upper member of Pioneer Formation is apparently the site of the Corregedor (=Cochise?) mine (loc. 14). The Geesaman fault in the vicinity has been intruded by Leatherwood granodiorite, and then reactivated. The shaft is at least 10 m deep and the extent of other workings is unknown. About 100 tons of ore containing 4 percent WO_3 were produced in the 1940's.

Nearby, many small workings are along the Geesaman fault. One is a mineralized inclusion of Abrigo Formation skarn in Leatherwood granodiorite, both sheared.

Workings along Geesaman Creek. Small workings along Geesaman Creek (loc. 146) have not been described previously. Those along the wash are south of the Geesaman fault in Paleozoic rocks, and show copper oxide along flexures in talc-bearing carbonate rocks. Those north of "5008" hill (Sanderson? deposit) are along a contact of Rice Peak porphyry (younger variety in part) with Oracle Granite.

Bluff mine group. These workings (loc. 147) in Paleozoic rocks above Alder Canyon were not previously thought to have any relation to the Geesaman fault, but plate 1 shows them to be along a window through a splay of that fault. Escabrosa Limestone is the main host rock; small intrusions of Leatherwood granodiorite are present nearby (Janecke, 1986), and wollastonite in Escabrosa against the Geesaman fault splay to the southeast suggests that Leatherwood-derived hot fluids traveled along its gouge.

The workings, consisting of tunnels and prospecting trenches, exploited galena, sphalerite, and copper sulfides as replacements in the limestone. About 80 tons of Pb-Zn-Cu-Ag-Au ore were produced in 1951-1957 (Keith, 1974).

Alder Creek placers. In Alder Creek near Bluff mine, placer gold was mined from alluvium between 1934 and 1940. About 22 oz. of gold were recovered (Johnson, 1972). The distribution of the placer workings mimics the distribution of the Geesaman fault splay exposed on both slopes above the stream; probably the fault controlled gold supply.

Marble Peak District

Mineral deposits of the Marble Peak district are contact-metamorphic, developed along the contact of the Leatherwood granodiorite intrusive sheet with overlying carbonate Paleozoic rocks of the Abrigo, Martin, and Escabrosa Formations. Individual ore bodies are rather small; those exposed at the surface (pl. 2B) or described in older literature (Peterson and Creasey, 1943) are all localized along pre-Leatherwood faults or other irregularities in the Leatherwood contact, and typically trend WNW.

Skarn ore may be strictly along the contact of sedimentary carbonate units and intrusive granodiorite, along a favorable horizon not in exposed contact with granodiorite, or may form "Christmas tree"-shaped bodies rooted on the intrusive contact. Between the Geesaman and Daily mines, mapping at 1:1200 by Peterson and Creasey (1943) showed a number of "tactite" bodies with Christmas-tree morphology, rooted on the granodiorite-marble contact (fig. 66). Some of these bodies are 60 m (200 ft) high and 120 m (400 ft) long.

Typical ore is garnet-epidote-diopside(-magnetite) skarn containing the sulfide minerals chalcopyrite, bornite, and chalcocite. Locally, galena, sphalerite, scheelite, molybdenite, and covellite occur. A little gold and silver are present. Sulfide minerals and epidote typically show retrograde relations relative to garnet and pyroxene.

The district had been worked for short periods in 1882 and 1910-1912 (Harrison, 1972), then intermittently since 1917, with most activity in the 1930's and 1940's. Total production prior to 1968 has been about 136,000 short tons of ore averaging about 3 percent copper, with much lesser amounts of lead, zinc, tungsten, gold, and silver, for a total value of $860,000 (Keith, 1974).

Current operations at the Oracle Ridge mine (near the former Geesaman mines) are exploiting primarily the copper and silver resources of the district, which are estimated as 8.4 million tons of ore with a grade of about 2.3% copper.

In spite of the sporadic mining activity, no detailed geologic map of the Marble Peak district had been publicly available. Peterson and Creasey (1943) presented only detailed maps on individual mine areas, and Braun (1969) did not subdivide the Paleozoic carbonate rocks. A network of drill roads make local mapping fairly easy at present, and plate 2B was therefore compiled at 1:12,000.

Geesaman mine group. These workings (loc. 148) are a focus of great expansion by present mining activities at Oracle Ridge mine. As mapped by Peterson and Creasey (1943), they follow an intrusive contact of Leatherwood granodiorite in marbleized Paleozoic limestone, striking N. 40 to 70° W., dipping about 40 to 50° SW, along which is developed a sulfide-rich skarn 3 to 10 m thick. At an elevation of about 5750 ft (1753 m) the Leatherwood intrusive sheet apparently jumps upsection to the southwest along this same strike, forming a corner of skarn ore. This may be either a fold or an intrusive feature; whether a pre-Leatherwood fault is present is unclear. Workings along this trend extend from elevations of about 5700 to 6000 ft (1740 to 1830 m) for distances of about 1000 ft (300 m).

My work on the ground surface near these workings shows that a 0.3 m garnet skarn with minor sulfide separates Leatherwood granodiorite and recrystallized Escabrosa Limestone. Rooted in this selvage are a few "Christmas trees" of skarn about 5 m high. Striking toward these workings from the west is a contact with Martin and/or Abrigo Formations (pl. 2B) associated with altered Leatherwood granodiorite, suggesting that the workings may be in basal Escabrosa, and may involve Martin and/or Abrigo in the subsurface.

The current (1981) topographic map shows the Geesaman mine group uphill to the west at an elevation of about 6300 feet (1920 m). These are apparently more recent workings than those originally called Geesaman (Geesman) mine, and are presently being worked as part of the Oracle Ridge mine. They may be along the same contact, judging from the appearance of ore being removed, but this is unclear because the adit is somewhat distant from the contact. At this elevation, Leatherwood granodiorite is in intrusive contact with Martin and/or Abrigo Formations rather than Escabrosa Limestone at the ground surface (pl. 2B).

Ore at the Geesaman mine, in addition to minerals produced throughout the district, contained covellite and a little scheelite. Production for the Geesaman mines has been listed together with that for the Daily mine, as the two mines were operated together from 1937 to 1940, a period of great productivity (Peterson and Creasey, 1943). More than half of the total production of 115,000 tons of ore can be attributed to the Geesaman mine, however. Copper grade was variously reported as 2 to 3 percent. One ton of high-grade tungsten ore was shipped.

Daily mine group. Workings described by Peterson and Creasey (1943) for the Daily mine (loc. 149) are along the intrusive contact between Leatherwood granodiorite and marbleized Paleozoic limestone, here along the northern limb of a small anticline in marble trending ENE. This fold is not reflected in the attitude of the intrusive contact which is nearly horizontal. Skarn about 3 to 5 m (10 to 15 ft) thick constituted sulfide ore along two anticlines and one syncline on the intrusive contact, which trend N. 60° W., and plunge gently west, spaced about 12 m (40 ft) apart. The trend of these folds is such that they *could* be the same structures as in the Geesaman mines. The workings were mostly stopes in an area of about 150 x 130 ft (45 x 40 m).

About 80 ft (25 m) above these workings, Peterson and Creasey (1943) mapped a subsidiary sill of granodiorite about 6 to 10 m (20 to 35 ft) thick. This sill is capped by skarn about 3 m (10 ft) thick, but skarn is absent below the sill (fig. 106).

Most of my observations were made at the workings labeled Daily mine group on the current (1981) topographic map, south about 400 m (1300 ft) from those described above. These workings follow a down-to-the-north step on the intrusive contact of Leatherwood granodiorite and recrystallized Escabrosa Limestone that strikes about N. 60° W., dipping steeply (pl. 2B). This step is associated with a steep Leatherwood dike. Leatherwood granodiorite itself is unfaulted; that is, any faulting predates Leatherwood intrusion. Garnet skarn is concentrated in the resulting corner of Escabrosa, and apparently constituted sulfide ore. Garnet rock also forms Christmas-tree bodies rooted in the intrusive contact (fig. 69), with greenish garnet forming the exterior parts of the bodies.

Ore of the Daily mine was typical of the district except for the presence of minor scheelite. Production has been included with the Geesaman mine. MRDS record M050641 reports intermittent production as recent as 1968.

Leatherwood mine group. Workings described by Peterson and Creasey (1943) are along an irregular northeast-dipping contact (pl. 1) of Leatherwood granodiorite and metamorphosed impure dolomitic rocks. About 700 ft (210 m) of workings in two main adits were seldom in ore. Many additional small workings are in magnesian skarn and dolomite, apparently of the Martin Formation (pl. 2B). To the northeast, the Martin is bounded by a fault into which is intruded a dike (not shown), apparently of Rice Peak porphyry's younger facies. Northeast of this fault, the sedimentary sequence has been dropped down 10 to 20 m (30 to 65 ft), but the intrusive Leatherwood contact steps up 10 to 20 m, so that Leatherwood is in contact with Escabrosa Limestone and the Martin is missing. This fault must therefore be pre-Leatherwood. Workings mapped by Peterson and Creasey apparently did not reach this structure. I suspect that skarn is present in this area because the fault produced a corner of favorable host rock.

Ore of the Leatherwood mines was typical for the district except for local high magnetite content. Exploitation concentrated on oxidized ore that when hand sorted showed copper grades of 5 to 12 percent. Production was intermittent from 1939 to 1968, and totaled about 3000 tons of ore (Keith, 1974).

Stratton mine. Approximately 70 m (240 ft) of workings described by Peterson and Creasey (1943) are in a 1-m zone in skarn among carbonate, spotted hornfels, and quartzite layers striking N. 35 to 45° W. and dipping 20 to 25° NE. These are probably Martin and Abrigo Formations (pl. 2B). Workings extend intermittently along strike for about 380 m (1250 ft), at the south end of which a northeast-striking left-lateral fault (not shown) is said to offset the zone several hundred feet. The main Leatherwood sheet is at least 20 m (65 ft) below the worked horizon.

Stratton mine ore contained a little molybdenite in addition to the typical skarn mineral assemblage. Reported production was only a few tons of high-grade ore. No road was ever completed to the workings.

Keith (1974) stated that the Stratton mine is the site of the Old Hat workings that gave their name to the entire mining region, as far north as Mammoth. However, Harrison (1972) and others ascribed this name to a former mine near the Magrugador workings (loc. 141).

Hartman-Homestake mine group. Workings here are the westernmost of the district (pl. 2B). Those shown by Peterson and Creasey (1943) total about 370 m (1200 ft), along units they designate as hornfels and schist about 60 m (200 ft) thick, underlying skarn and marble, striking about N. 60 to 70° E. and dipping about 30° SE. These seem to be Escabrosa, Martin and Abrigo Formations down to the Southern Belle Member, which is the unit intruded by Leatherwood granodiorite at the ground surface (at elevations of about 6500 ft or 1980 m).

My mapping shows that several steep faults striking about N. 60° W offset the base of the Escabrosa Limestone down to the north for a total of about 100 m (300 ft). Locally, Leatherwood dikes intrude these faults. Some of the workings are along these structures and bedding-plane faults striking northeast along the Martin-Abrigo contact. Ore taken from the fractures typically contained about 3% Cu, 5 to 7% Pb, 5 to 9% Zn, and a little silver (Peterson and Creasey, 1943), as chalcopyrite, sphalerite, and galena. Production was about 600 tons of ore containing Cu (3%) - Zn (3%) - Pb (5%) - Ag (30 oz./ton) in 1947 and 1948 (i.e. post-dating description by Peterson and Creasey, 1943), according to Keith (1974).

Related deposits. The Iron Cap group of claims along Alder Creek near Alder Box Spring (pl. 1) at the foot of Marble Peak are apparently along the bottom of the Leatherwood intrusive sheet. They are probably skarn deposits but little is known about them.

In Atchley Canyon (loc. 150) are workings in Abrigo Formation along a steep dike of porphyritic Leatherwood granodiorite. Epidosite is present but no mineralization was seen. No description of workings or record of production was encountered.

Knagge Mine

Along the Knagge Trail west of Kellogg Mountain, at about 6900 feet (2100 m) elevation, is the Knagge mine (loc. 92). It is along a shear zone striking N. 60° E., dipping about 60° southeast, in the Knagge (garnet-biotite) granitic pluton. Along strike this shear zone cuts a large elongated metadiabase xenolith. To the northeast the Knagge granite intrudes the Kellogg and Knagge faults, separating Apache Group rocks from Abrigo Formation (pl. 1).

The workings at Knagge mine apparently have not been described. The visible mineralization is copper oxide and molybdenite. The mine was probably worked in the 1940's. The name is reported by local residents.

Other Workings

Range crest. Several prospects along the range crest are small copper-bearing skarn deposits in Abrigo Formation where it has been intruded by Leatherwood granodiorite (MRDS records M051053 to 051055). Probably the most extensive such workings are those of the Zimmerman deposit (loc. 151). Epidote-garnet skarn about 2 m thick contains copper oxide in Abrigo Formation dipping gently southeast. An adit and several prospects probably date from early in this century. No production figures are published. Across upper Sabino valley from these workings, the same contact of Abrigo above intrusive Leatherwood is well exposed in road cuts (loc. 73), and shows extensive garnet-epidote alteration but little mineralization.

South-facing main range. A line of prospects follows the contact of Catnip Canyon banded granite with the overlying inclusion zone (pl. 1), from Molino campground west past the prison camp. At least seven prospects are present, along quartz veins that are subhorizontal or gently north-dipping. Wallrocks show incipient mylonitization, and the quartz itself is locally sheared. Visible mineralization consists of copper oxide. Heylmun (1989) reported visible gold at one of the prospects, which he referred to as the Valerie May prospect. Hinkle and others (1981) found elevated zinc contents in sediment samples from this area.

Forerange. The Catalina forerange contains very few showings of mineralization in the study area. One of the most substantial known (MRDS record M050689) is in lower Bear Canyon below Seven Falls, and consists of a quartz vein containing epidote and copper oxide, striking across the canyon.

Along the Catalina detachment fault and its footwall near Sabino Canyon at the southwest corner of the study area are prospect pits (Ryan, 1982). No mineralization was seen or reported. Hinkle and others (1981) reported detrital gahnite nearby.

Along the Catalina detachment fault west of the map area is the Pontatoc mine, which produced 5000 tons of ore containing 4 percent copper and a little gold and silver in 1907 to 1917 (Keith, 1974). The deposit is thought to be detachment-related and its age to be mid-Tertiary.

MINERAL RESOURCE POTENTIAL

Most of the districts described above have potential for undiscovered deposits or extensions in little-explored areas. Such potential is described in this section, as well as potential for deposit types not represented by known deposits. Potential for metallic deposits is discussed by region, whereas that for non-metallic deposits is by commodity.

Metallic Deposits

San Manuel district and related deposits. Interpretation of the San Manuel-Kalamazoo deposit as an envelope developed along a tabular igneous body (fig. 105) has some implications for distal ends of the deposit. For example, mineralization is known in parts of the deposit east of the Cholla fault (Schwartz, 1953), but the bullet-shaped termination of ore in the model of Lowell (1968) and Lowell and Guilbert (1970) would imply that the amount of ore there is small. In contrast, linear extrapolation of the converging upper and lower bounds of the ore envelope (fig. 105) across the Cholla fault would imply the presence of a few percent of the San Manuel orebody east of the Cholla fault. Files in the Anaconda collection imply 1.54 million tons of ore at 1.08% Cu east of the fault (letter from Art Barber to Roland Mulchay, May 19, 1961). On the other end of the tabular envelope, ore could be present southwest of the Red Rock fault, but its burial depth is likely to be over a kilometer there.

Interpretation of the San Manuel porphyry in the broad sense as emplaced along an unconformity between Oracle Granite and American Flag Formation suggests that this unconformity toward the SSE could contain additional deposits. Most of this area is concealed by San Manuel and Quiburis Formations. The most favorable situation would probably be the intersection of the unconformity with porphyry dikes oriented parallel to the San Manuel body.

The relation of San Manuel mineralization to that in Laramide volcanics across the San Pedro valley is unclear.

Across the Mogul fault, the prospect in Nugget Canyon (loc. 146) may possibly be related to a San Manuel-like system. It occurs in American Flag Formation above a faulted unconformity that in this area is on Paleozoic rocks.

Mammoth district. Interpretation of the Mammoth district as related to the Cloudburst-Turtle detachment system suggests that the Cloudburst and Turtle faults themselves should be mineralized. Veins of Mammoth morphology are present in the Turtle fault at Tucson Wash (loc. 132), and the Cloudburst detachment fault is apparently mineralized at Tar mine.

Some authors have suggested that the junctions of detachment faults with their steep side ramps are favorable for detachment-related mineralization (for example, Long, 1992). This is consistent with the distribution of mineral assemblages (described by Peterson, 1938) in the Mammoth vein set, reconstructed by Force and Cox (1992) to show concentric zonation outward from a point that would be under the intersection of the plane of the vein set with the intersection of the Cloudburst and Turtle faults.

Possibly detachment-related alteration and mineralization as at Shultz Spring may have potential for additional deposits (Force and Cox, 1992). The relation to detachment or rhyolite is not clear at either the Shultz Spring or the probably correlative Ford mine across the San Manuel fault, and exposed mineralization is weak.

Mogul fault-related districts. The Campo Bonito district might be thought to have large potential because the Mogul is such a large fault, but favorable host rocks are far more restricted. The best deposits of the district are bounded by faults, which outline a small block of Escabrosa Limestone on all four sides in map view and at shallow depth below. This block of Escabrosa is unique in the immediate area, but limestone is in contact with the Mogul fault at the 3C Ranch and probably is in contact at depth nearby.

In the Southern Belle mine area, the unconformities separating Dripping Spring Quartzite, the Campo Bonito unit, and Bolsa Quartzite, and deformed paleosols within the Dripping Spring, may be favorable horizons that could be explored in surrounding areas. The significance of mineralization in this narrow stratigraphic interval may be the oxidizing capacity of the Campo Bonito unit (fig. 107).

Resource potential in the American Flag district is probably small, but ore controls are not well understood. Possibly the greatest potential is in the largest fault structure, that cutting the thick diabase dike at American Flag on the southwest at a small angle (pl. 1). Quartz-siderite veins are well developed along this zone.

Remainder of Geesaman-Mogul block. With the exception of the Southern Belle district and the Nugget Canyon prospect, both discussed in other contexts above, this block contains few mineral deposits and few promising prospects. Possibly such scarcity reflects the basement-cored uplift and laccolithic intrusions that characterize most of the block.

Many prospects of this block are hosted by the Escabrosa, Abrigo, Campo Bonito, and middle Pioneer horizons that form stratigraphic-geochemical barriers (fig. 107), and several can be related to faulting possibly synchronous with mineralization. Potential for further mineralization nearby probably is a function of the same variables. Proximity of Rice Peak porphyry appears to be a poor predictor.

At the mouth of Stratton Canyon, the juxtaposition of the Stratton detachment fault with favorable older features in the footwall (vent facies of Rice Peak porphyry and associated intruded fractures) may give this area a potential for mid-Tertiary mineralization.

The lacustrine facies of Dripping Spring Quartzite, present in this block in the Canada del Oro area, is similar to that containing uranium in the Sierra Ancha area about 100 km to the north. This facies is also present in Atchley Canyon south of the Geesaman fault (pl. 1B).

The Campo Bonito unit is characteristically enriched in iron (table 4), present as hematite cement. However, iron content is too low to be considered a resource.

Geesaman fault area. Deposits along the Geesaman fault constitute a newly recognized population that is finely zoned from aplites and greisen on the west through tungsten deposits to base-and precious-metal deposits on the east. This zoning implies a predictability that could spur further exploration. Analogies between the Mogul and Geesaman faults also suggests more mineral potential along the Geesaman than has been realized thus far. The presence of greisen suggests that tin potential be considered.

Where the Geesaman fault abuts the Marble Peak district, Tertiary remobilization of the older mineralization may have resulted, except where the fault plane was sealed by Leatherwood intrusion. Hot fluids along the Geesaman fault in the Alder Creek window formed local wollastonite, and Laramide skarn deposits may have formed there also.

Marble Peak district. Localization of the sulfide-rich skarn bodies by individual pre-Leatherwood folds and faults, and by other irregularities in the shape of the intrusive contact, makes prediction of deposit formation difficult. Varying host rocks and distance from intrusion make prediction of mineral assemblages difficult also. However, northwest-striking steep faults such as those at Leatherwood and Hartman-Homestake mine groups may continue as intruded faults for some distance (pl. 2B) in the subsurface and impart some potential.

The known skarn orebodies are above the Leatherwood sheet, and examination of the same sedimentary horizons below the sheet near Geesaman and Alder Creeks has revealed thin skarns in only a few localities. Similar asymmetry across thin Leatherwood sills (fig. 106) was observed by Peterson and Creasey (1943), suggesting that the potential for skarn deposits below the intrusive sheet is less than that above it. However, intersections of the most favorable horizons with the base of the Leatherwood sheet should be examined.

Elsewhere south of Geesaman fault. Between the Geesaman fault and the range crest are large areas that are difficult of access and still poorly known. Exploration has probably been cursory. However, I find that this area contains large areas of Abrigo Formation (pl. 1), where previous published maps showed none. The Abrigo is a favored host for mineralization, partly because it forms the lowest carbonate unit in many areas (fig. 107), partly because it overlies Bolsa Quartzite that locally functions as an aquifer, and partly because its finely laminated impure nature makes a variety of reactants available. Particularly where the Abrigo is intruded by Leatherwood sills and metamorphosed to calc-silicate skarn, sulfide deposits may occur. Copper liberated by metamorphism of Proterozoic diabase (Force, in press) may be trapped by Abrigo.

The geology of the Loma Linda-Radio Ridge-Carter Canyon area of Mt. Lemmon is similar to that of Marble Peak, and where pre-Leatherwood faults occur, there is a potential for skarn deposits. A few such deposits occur, as at locality 152. Skarns exposed on the Mt. Lemmon highway near Loma Linda are well-developed but sulfide-poor.

Except along the Catalina detachment fault, areas south of the range crest have no obvious potential for metallic mineral deposits, because of unreactive rocks and great paleo-depths during both the Laramide and mid-Tertiary. Hinkle and Ryan (1982) report an area in which sediment samples contain gahnite and scheelite, and anomalous base metal contents, but no indication of mineralization was found. The rusty-weathering zone at the base of the Wilderness granite sill may be worthy of investigation, but no sulfide minerals have yet been found in it. The Catalina detachment clearly has potential for detachment-related deposits, but the only reported occurrences of this type are outside the study area.

Non-metallic Deposits

Marble. White marble, mostly after Escabrosa Limestone, is common on Marble Peak, where it locally shows attractive folds in bedding. Below the Leatherwood sheet in Geesaman Creek (loc. 41) the same rock is more easily accessible. Joints are 0.5 m or more apart, allowing some dimension-stone uses.

Slate. Slate was encountered in Pioneer Formation in several places but appears to be of the best quality just above Juanito Spring (fig. 20, loc. 11). At this locality both purple slate with green blebs and gray slate occur. Total slate thickness is estimated as 30 m (100 ft).

Pegmatite minerals. Feldspar, mica, and quartz are available as coarse grains from Lemmon Rock pegmatite-aplite near the range crest. Feldspar is perthitic, commonly intergrown with quartz in graphic textures. Mica, including both muscovite and biotite, is deformed. Garnet is locally coarse, but with the clarity of common garnet. The Lemmon Rock pegmatites are poor in valuable accessory minerals.

Emery and other abrasives. Skarn deposits at Leatherwood mine locally consist of garnet and magnetite with lesser epidote, quartz, and other minerals. This could form an acceptable abrasive mixture.

Wollastonite. Wollastonite occurs with other minerals in skarns as at Marble Peak, and is found disseminated without other calc-silicate minerals mostly in Escabrosa Limestone at Marble Peak and the Alder Creek window. It has not been found in pure masses.

Appendix 2
Locality Descriptions

(shown on plate 1b)

1. Pinal Schist in the gorge of Alder Creek (Mt. Bigelow quadrangle). The schist here is pelitic and shows typical crenulations of quartz segregations (fig. 11).

2. Pinal Schist intruded by Oracle Granite in the bed of Edgar Canyon south fork (Mt. Bigelow quadrangle). One foliation in Pinal is cut by Oracle, but another is shared with Oracle and with dikes of porphyritic Leatherwood granodiorite. Nearby Knagge pluton shows no foliation.

3. Unconformity between Pinal Schist and overlying Scanlan Conglomerate Member of Pioneer Formation on the southern slopes of Kellogg Mountain (Mt. Bigelow quadrangle). The Scanlan here is a stretched-pebble conglomerate (fig. 12) with fabric of a typical Laramide orientation. In the Pinal this fabric is overprinted on quartz segregations. Downslope, Pinal is apparently intruded by Oracle Granite, then by unfoliated granite at the top of the Wilderness sill. Upslope are Pioneer Formation and Dripping Spring Quartzite with local Barnes Conglomerate Member at the base.

4. Deformed xenoliths of Oracle Granite in Wilderness Granite (fig. 15), eastern part of Green Mountain (Mt. Bigelow quadrangle). Laramide fabric of Oracle suggests top to northeast but is cut by leucogranite and pegmatite bearing garnet.

5. Unconformity between underlying Oracle Granite and overlying Scanlan Conglomerate Member (fig. 16B) along Sunset Trail (Mt. Bigelow quadrangle). The unconformity has been intruded here by a thin sill of Leatherwood granodiorite. Just downslope is intrusive Lemmon Rock pegmatite-aplite at the top of the huge Wilderness Granite sill. Oracle Granite here is badly deformed and muscovitic, as a result of some combination of Precambrian soil formation, Laramide deformation, and Eocene metasomatism. The Scanlan has been stretched, silicified, and locally feldspathised so that its texture is variably obscured.

6. Unconformity between underlying Oracle Granite and overlying Scanlan Conglomerate Member (fig. 16A), west flank of Oracle Ridge (Oracle quadrangle). The unconformity here is little-deformed and sharp. This exposure and overlying sandstones form a northeast-dipping screen among Rice Peak porphyry intrusives and a fault.

7. Unconformity between underlying Pinal Schist and overlying Scanlan Conglomerate Member (fig. 17) in the bed of Canada del Oro (Oracle quadrangle). The basal lithology of the Scanlan here is cross-bedded pebbly sand overlying a possible paleosoil. The conglomerate contains quartz pebbles.

8. Unconformity between underlying Oracle Granite and overlying Scanlan Conglomerate Member near Bat Well on Geesaman Wash (Mt. Bigelow quadrangle). A deformed granular phyllite up to 3 m thick at the base of the Scanlan probably represents a slightly transported paleosoil.

9. Sandy middle member of Pioneer Formation (fig. 18) in Atchley Canyon (Mt. Bigelow quadrangle). Cross beds indicate transport from the east.

10. Contact of Barnes Conglomerate Member of Dripping Spring Quartzite and underlying Pioneer Formation, in streambed below Mary West mine (Campo Bonito quadrangle). The base of the Barnes is sandy here. The Pioneer is a laminated to cross-laminated olive to maroon sandy siltstone (figs. 19, 24).

11. Slaty upper member of the Pioneer Formation near Juanito Spring (Mt. Bigelow quadrangle). The slaty lithologies include purple, gray-green, and spotted pelitic to silty rocks, the latter showing cross-lamination (fig. 20) or graded beds.

12. Diabase dike cutting contact of Pioneer Formation pyroclastics and Barnes Conglomerate Member of Dripping Spring Quartzite (fig. 21) on Control Road (Mt. Bigelow quadrangle). A pinchout of Barnes here is complicated by a Rice Peak porphyry dike and related fault.

13. Pyroclastic facies of Pioneer Formation cut by dike of meta-diabase, Edgar Canyon (Mt. Bigelow quadrangle). Both lithologies are strongly deformed by a typical Laramide fabric. Lacustrine facies of Dripping Spring Quartzite is just downstream.

14. Coarse pyroclastics in Pioneer Formation (fig. 22), Corregedor mine area (Mt. Bigelow quadrangle). Rolled quartz eyes pseudomorph beta-quartz phenocrysts.

15. Coarse pyroclastics in Pioneer Formation on Control Road (Mt. Bigelow quadrangle). Flattened pumice suggests ash-flow origin. See table 2 for composition.

16. Upper member of Pioneer Formation on Oracle Ridge (Oracle quadrangle). Cleaved silty lithology is apparently not pyroclastic but is potassic (table 2). A large quartz vein shows chlorite-pyrite margins.

17. Phyllite and schist after Pioneer Formation at Inspiration Rock (Mt. Lemmon quadrangle). The upper member of Pioneer Formation here consists of phyllite and schist that has been deformed at least twice, such that early cleavages are folded. The western end of the outcrop consists of Lemmon Rock pegmatite-aplite. Under a cliff runs the garnet aplite band of localities 86 and 87.

18. Barnes Conglomerate Member in the bed of Peppersauce Wash (Campo Bonito quadrangle). Barnes contains pebbles of quartz, quartzite, and jasper in a maroon sandy matrix. The base is here against feldspathic sand. The top is a sharp contact with quartzite. This is the most accessible outcrop of little-deformed Barnes in the study area.

19. Lacustrine facies of Dripping Spring Quartzite in the bed of Canada del Oro (Oracle quadrangle). The upper member of Dripping Spring Quartzite here consists of pyritic laminated dark argillite.

20. Lacustrine facies of Dripping Spring Quartzite overlain by diamictite facies of Campo Bonito Formation in Atchley Canyon (Mt. Bigelow quadrangle). The upper member of the Dripping Spring consists of laminated dark argillite, locally slaty, here and upslope to the south.

The diamictite rests directly on diabase intruded into the Dripping Spring. The clasts are mostly of quartz, exceedingly angular at the base, in a poorly sorted, polymineralic dark matrix.

21. Bear Wallow thrust at Bear Wallow (Mt. Bigelow quadrangle). The thrust places Dripping Spring Quartzite, locally thrown into recumbent isoclinal folds, over finely bedded marble of Mescal Limestone (fig. 24). Gouge about 1 m thick shows Z-shaped folds (fig. 89). One dike of pegmatite is cut by the thrust, but another cuts it.

22. Mescal Limestone in core of Oro syncline (Oracle quadrangle). Mescal here is cherty limestone and marble. Its base has been altered to epidote and copper minerals. This is the southernmost outcrop area of Mescal in the syncline (pl. 1).

23. Fresh Proterozoic diabase on road north of American Flag Hill (Campo Bonito quadrangle). Diabase forms a thick "dike" (see text) in Oracle Granite and related aplite. The central portions are rather coarse. Except for retrograded shear zones, this is a single intact body of diabase which locally preserves the freshest mineral assemblages found in diabase of the study area. The minerals include olivine, clinopyroxene, partly as oikocrysts, plagioclase, and skeletal magnetite (fig. 28).

24. Proterozoic diabase on trail above Peppersauce Wash (Campo Bonito quadrangle). Diabase here forms a thick sill in Pioneer Formation. See table 3 for composition.

25. Biotite-amphibole-plagioclase schist after diabase, Butterfly Trail (Mt. Bigelow quadrangle). Relict subophitic texture and chill zones can still be recognized in some intervals. This sill has intruded Dripping Spring Quartzite.

26. Proterozoic diabase in Purcell window (Mammoth quadrangle). This "dike" is in Oracle Granite but is quite close to the unconformity with Cloudburst Formation (up-gully), so the angular relation between them is clear. See table 3 for composition.

27. Proterozoic diabase at Highjinks mine (Campo Bonito quadrangle. The margins of a diabase "dike" in Oracle Granite are the locus of minor gold deposition in quartz. See table 3 for diabase composition.

28. Proterozoic diabase in Oracle Granite (Oracle quadrangle). Table 3 shows the composition of one of several diabase "dikes."

29. Campo Bonito Formation maroon siltstone below Southern Belle mine (Campo Bonito quadrangle). Massive siltstone passes upward to more laminated siltstone (fig. 29) with bleached spots. Textures are poorly sorted with angular grains and a matrix entirely of iron oxide (fig. 31). See table 4 for composition. The top and base of the formation here are both unconformities. The base is on Dripping Spring Quartzite

containing a diabase dike. The top is overlain by Bolsa Quartzite (fig. 29).

30. Campo Bonito Formation maroon siltstone in Peppersauce Wash (Campo Bonito quadrangle). Massive siltstone alternates with intraformational conglomerate and contorted sandy beds. The texture is poorly sorted with angular grains and matrix mostly of iron oxide (fig. 31). See table 4 for composition.

31. Campo Bonito Formation maroon siltstone above Southern Belle mine (Campo Bonito quadrangle). In this condensed section of Campo Bonito, basal diamictite and upper laminated and bleached sandy siltstone comprise the entire formation.

32. Campo Bonito Formation diamictite in the bed of Canada del Oro (Oracle quadrangle). The top but not the base of this channel-fill is well exposed here. The paleovalley walls were apparently steep (fig. 29). Diamictite fill fines upward and is poorly sorted and angular, with clasts of sedimentary, volcanic, plutonic, and foliated metamorphic rocks (fig. 33). Hematite replaces labile pyroxene, sphene, biotite, etc.

33. Campo Bonito Formation diamictite in the bed of Canada del Oro (Oracle quadrangle). A section through diamictite including the top but not the base is exposed here. The basal exposures are coarse massive diamictite, which grades up through finer diamictite into pebbly varved poorly sorted dark sandstone.

34. Bolsa Quartzite in Atchley Canyon (Mt. Bigelow quadrangle). A well-exposed typical transition (fig. 35A) passes upward from lower cross-bedded gritty sandstones with dark maroon to gray argillite partings, into better sorted white to pink sandstone with small herringbone cross-beds. The Bolsa here rests on Proterozoic diabase with no indication of a paleosoil.

35. Bolsa Quartzite near Butterfly Peak (Mt. Bigelow quadrangle). The Bolsa rests unconformably on altered marble of Mescal Limestone and latered diabase, and is overlain by Abrigo Formation baked to epidote-chlorite calc-silicate rock but preserving flaser bedding (fig. 37).

36. Upper intrusive contact of Lemmon Rock pegmatite-aplite at Lemmon Rock (Mt. Lemmon quadrangle). Typical pegmatite contains skeletal biotite and xenoliths of Leatherwood granodiorite (fig. 61). It is intruded into schist and phyllite of Pioneer Formation and Proterozoic diabase, locally showing garnet-epidote assemblages ascribed to Leatherwood intrusion. On the slopes just above is a poorly exposed unconformity between this sequence and overlying Bolsa Quartzite. That is, the sub-Bolsa unconformity here cuts out most of the Apache Group.

37. Abrigo Formation in Alder Canyon (Mt. Bigelow quadrangle). Finely bedded Abrigo here is slightly epidotic but otherwise unmetamorphosed, and has not suffered carbonate-destructive diagenesis.

38. Lower (Three C) member of Abrigo Formation in the bed of Peppersauce Wash (Campo Bonito quadrangle). Abrigo here has been leached of carbonate components and is therefore somewhat vuggy, earthy, and soft even in waterworn exposures. Diagenetic removal of carbonates may also have flattened the flaser structures of bedding.

39. Type section of Mt. Lemmon unit on road to Mt. Lemmon (Mt. Lemmon quadrangle). The base of the unit at the northern end of the exposure is deformed but apparently cut into Abrigo Formation calc-silicate rock and marble. A basal conglomerate is exposed at some contacts. The fresh sandstone adjacent to the base contains tiny andalusite porphyroblasts. Near the mapped locality, slightly weathered sandstone is poorly sorted and micaceous, shows cut-and-fill structures, and has many sandy phyllite interbeds. Sandstones are amalgamated such that interbeds are discontinuous. Calc-silicate knots suggest former calcareous nodules.

40. Mt. Lemmon unit on Red Ridge (Mt. Lemmon quadrangle). The base of the unit rests on the middle (Southern Belle) quartzite member of Abrigo Formation along this ridge. The Mt. Lemmon is less indurated and more poorly sorted. Staurolite and andalusite are present in more pelitic lithologies.

41. Marble after Escabrosa Limestone near Geesaman Wash (Mt. Bigelow quadrangle). Escabrosa has been baked here by intrusion of the overlying Leatherwood granodiorite sheet to coarse white marble with widely spaced joints (fig. 38, 44).

42. Lower Horquilla Limestone on Marble Peak (Mt. Bigelow quadrangle). Anastomosing dolomite-limestone alternations (fig. 40A), apparently after flaser bedding, locally contain fossils.

43. Upper Horquilla Limestone on Marble Peak (Mt. Bigelow quadrangle). Finely bedded dark limestone here contains zoned dolomite-chert-minor phosphate nodules (fig. 40B).

44. Earp Formation on Marble Peak (Mt. Bigelow quadrangle). Argillite, quartzite, and marble are interfingered or interfolded here (fig. 42).

45. Earp Formation near Geesaman Wash (Mt. Bigelow quadrangle). The "jelly-bean" (chert-pebble) conglomerate is present below quartzite. It has a calcareous matrix but contains chert nodules in addition to the rounded chert pebbles.

46. Lower member of American Flag Formation at Nugget Canyon (Campo Bonito quadrangle). Limestone-pebble and -cobble conglomerate, locally in overturned beds, lies below Paleozoic carbonate units along reverse faults. The conglomerate beds are commonly cleaved. A dike of Rice Peak porphyry younger variety cuts the American Flag.

47. Lower member of American Flag Formation in valley south of Peppersauce Wash (Campo Bonito quadrangle). Limestone-cobble conglomerate is structurally below brecciated Escabrosa Limestone. A sill of Rice Peak porphyry younger variety separates the two units.

48. Middle member of American Flag Formation on Control Road (Campo Bonito quadrangle). Gritty feldspathic conglomerate, greenish and very tough due to volcaniclastic matrix, alternates with conglomerate containing quartzite and granite pebbles derived from Precambrian rocks.

49. Tuff in middle member of American Flag Formation (Campo Bonito quadrangle). A whitish vitric tuff about 1.5 m thick occurs among outcrops of conglomerate with quartzite cobbles and greenish feldspathic grit.

50. Unconformity at the base of the middle member of American Flag Formation (Campo Bonito quadrangle). Conglomeratic middle member unconformably overlies cleaved, steeply dipping lower member (here containing epidotic limestone pebbles). The outcrop of Escabrosa Limestone here is a rotated block in lower member.

51. Plant fossils in middle member of American Flag Formation in streambed (Campo Bonito quadrangle). Pebbly feldspathic sandstone (somewhat weathered) contains plant stem fragments. Just downstream is an apparently intrusive contact with Rice Peak porphyry (indeterminate type).

52. Sill of Rice Peak porphyry in middle member American Flag Formation (Campo Bonito quadrangle). A sill about 3 m thick of Rice Peak porphyry younger variety is present in the middle member of the American Flag about 10 m above the unconformity on Escabrosa Limestone. This porphyry is the trachitic-textured plagioclase-amphibole-biotite rock (FT 820) for which U-Th dating on zircon was attempted with inconclusive results (Unruh, this volume). See table 7 for composition.

53. Upper member of American Flag Formation (Campo Bonito quadrangle). An unusual bed of conglomerate with welded tuff clasts is enclosed by more typical conglomerate beds with clasts of porphyry, apparently Rice Peak porphyry older variety.

54. Margin of American Flag Formation in northern area near Southern Belle wash (Campo Bonito quadrangle). Shattered conglomerate containing limestone and volcanic clasts (poorly exposed) is structurally overlain by shattered Escabrosa Limestone along a gently SW-dipping fault.

55. Probable American Flag Formation and San Manuel porphyry in south fork Mammoth Wash (Mammoth quadrangle). Conglomerate containing clasts of Oracle Granite and overlying volcanic rock of intermediate composition are intruded along the eastern side of the outcrop by an irregular NNW-trending dike of biotite-plagioclase porphyry. This is apparently San Manuel porphyry type B, and Unruh (this volume) obtained a U-Pb age of 67.8 Ma on zircon from FT 62 at this locality. The San Manuel fault separates this locality from the San Manuel Formation to the west.

56. Conglomerate of Cloudburst Formation lower member at waterfall in Cloudburst Wash (Mammoth quadrangle). Conglomerate here is highly indurated with silica, epidote, and chlorite due to alteration related to the Mammoth vein set (just above waterfall) and to baking related to both underlying Cloudburst detachment fault and intrusive rhyolite. Conglomerate contains clasts of Oracle Granite and some vesicular volcanic rocks that may have originated as bombs and lapilli. Downstream (upsection) are cross-bedded sandstone and volcanic rocks of intermediate composition. Table 5 lists compositions of both.

57. Unconformity of Cloudburst Formation lower member on Oracle Granite, Collins open cut (Mammoth quadrangle). Cloudburst Formation here is conglomerate containing cobbles of Oracle Granite and volcanic rocks, and rests unconformably on Oracle Granite (most clearly north of the open cut). Rhyolite

locally obscures the relation. Induration is moderate; epidote is apparently absent here. Barite veins occupy joints. This sequence is juxtaposed across the Mammoth "vein-fault" against Cloudburst Formation (except at the south end). See table 5 for composition of conglomerate.

58. Type section of Stratton unit from Control Road toward Stratton Wash (Mt. Bigelow quadrangle). Porphyritic gray volcanic rocks, locally flow-banded, are exposed on the road. Underlying these in intermittent outcrops to the south are vesicular microlitic volcanic rocks, then green to red conglomerate with gritty feldspathic matrix and poorly rounded pebbles of volcanic rocks and feldspathic sandstone, probably derived from American Flag Formation.

59. Stratton unit and Mogul fault in Peppersauce Wash (Campo Bonito quadrangle). Shattered reddish conglomerate of Stratton unit juxtaposed across Mogul fault (neo-Mogul of this report) against Quiburis Formation gravel to the southwest. Younger gravel then overlies this package. Stratton conglomerate here contain angular volcanic clasts and a possibly pyroclastic matrix. Locally epidote-chlorite-quartz veins and alteration occur in conglomerate.

60. Mylonitic Leatherwood granodiorite in Molino Basin (Agua Caliente Hill quadrangle). Leatherwood here forms an intrusive sheet 50 to 100 m thick in Oracle Granite. It has itself been intruded by leucogranite and pegmatite and then mylonitized (fig. 43). The north end of the outcrop shows a cataclastic overprint on mylonite (fig. 79).

61. Mylonitic Leatherwood granodiorite near Sycamore Reservoir (Agua Caliente Hill quadrangle). A small body of Leatherwood here is juxtaposed against the Gibbon Mountain leucogranite sill across the Romero Pass fault zone.

62. Xenoliths of dark granite in inclusion zone north of Molino Basin (Agua Caliente quadrangle). These xenoliths are apparently a variant of Oracle Granite. The xenoliths themselves contain boudins of Leatherwood granodiorite that line up in a manner suggestive of former Leatherwood dikes. The host rocks are various phases of Wilderness suite leucogranites. All the rocks are somewhat mylonitic.

63. Termination of Leatherwood granodiorite sheet in Alder Creek along Crystal Spring trail (Mt. Bigelow quadrangle). Massive Leatherwood passes southeastward into Dripping Spring Quartzite, via a 60-m interval in which porphyritic Leatherwood sills about 1 m thick intrude the quartzite, at nearly right angles to the contact itself.

64. Leatherwood granodiorite intrusive across Geesaman fault, Control Road (Mt. Bigelow quadrangle). Equigranular chilled Leatherwood is unsheared, in intrusive contact with Pioneer Formation in the northern wall of the Geesaman fault (fig. 87). A dike of Rice Peak porphyry younger variety is present.

65. Leatherwood granodiorite sill in gouge of Geesaman fault splay near Alder Canyon (Mt. Bigelow quadrangle). A little-sheared Leatherwood sill 1 to 3 m thick intrudes gouge 5 to 10 m thick of a gently dipping splay of the Geesaman fault. Below this splay is Pinal Schist, shattered and smeared near the fault. Above it is Escabrosa Limestone.

66. Leatherwood granodiorite from Control Road (Mt. Bigelow quadrangle). Fresh little-sheared Leatherwood with few Wilderness suite dikelets. See table 6 for composition.

67. Rice Peak porphyry older variety near Oracle Ridge (Campo Bonito quadrangle). Rice Peak, locally massive and fresh, forms a thick sill in Pioneer Formation as plagioclase-amphibole-biotite porphyry with epidote-chlorite-calcite alteration. See table 7 for composition. Unruh (this volume) obtained a U-Th date of about 71$\pm$8 Ma on zircon from sample FT BH here.

68. Two varieties of Rice Peak porphyry on dirt track east of Rice Peak (Campo Bonito quadrangle). A darker porphyritic variety occurs as inclusions in younger lighter-colored more equigranular porphyry with a trachytic texture. Both rocks are quartz-poor and altered to chlorite-epidote-sericite.

69. Deformed Rice Peak porphyry on Control Road (Mt. Bigelow quadrangle). Porphyry probably of the older variety has been flattened to a phyllitic rock with feldspar phenocrysts showing euhedral outlines perpendicular to foliation but smeared along foliation.

70. Two varieties of Rice Peak porphyry near Control Road and Nugget Canyon (Campo Bonito quadrangle). Middle (Southern Belle) quartzite member of Abrigo Formation has been intruded by foliated older variety of Rice Peak porphyry to the south, then faulted. Younger variety of Rice Peak appears to cut all these features and is not foliated.

71. Vent facies of Rice Peak porphyry near Control Road and Stratton Canyon (Campo Bonito quadrangle). Volcanic breccia containing porphyry clasts in volcanic groundmass grades into massive Rice Peak porphyry, both hosted by Escabrosa Limestone.

72. San Manuel porphyry in Purcell window (Mammoth quadrangle). Porphyry type B here forms a NE-dipping dike about 400 m thick in Oracle Granite. This locality exposes the freshest San Manuel porphyry known in surface exposures; alteration is minor but chlorite is present. The primary phenocryst mineralogy is plagioclase-biotite. See table 7 for composition. Unruh (this volume) obtained a U-Pb date of 68.1 Ma on zircon from sample FT 84 from this locality, and Jon Hagstrum used sites near here for three of four paleomagnetic investigation of San Manuel porphyry (Force and others, 1995).

73. Lemmon Rock pegmatite-aplite intruded in Leatherwood granodiorite, ski road (Mt. Lemmon quadrangle). Dikelets of pegmatite and aplite, locally with garnet (fig. 65), form a near-vertical array in Leatherwood host rock (fig. 51). These dikes are structurally just above the main mass of pegmatite-aplite. The western boundary of these outcrops is an intrusive contact of Leatherwood in Abrigo Formation, along which garnet-epidote skarn is found.

74. Leucogranite of Wilderness suite intrusive into foliated Leatherwood granodiorite, waterfall on Butterfly Trail (Mt. Bigelow quadrangle). Leucogranite and pegmatite dikes are unfoliated and undeformed, in foliated Leatherwood host rock (fig. 52), here as a large porphyritic sill.

75. Seven Falls sill in Sabino Canyon (Sabino Canyon quadrangle). Biotite leucogranite of this sill is exposed in the canyon at the crest of the forerange anticline, below outcrops of deformed Oracle Granite (fig. 53). The strong penetrative deformation in leucogranite is ductile in all constituent minerals, suggesting the locality is at the structural base of mylonite. A thin dike of little-deformed pegmatite is also present.

76. Seven Falls sill and Bear Canyon fault at Seven Falls, Bear Canyon (Sabino Canyon quadrangle). This locality repeats the features of locality 75. The steep Bear Canyon fault forms the southeastern margin of the leucogranite, and is marked by red gouge and breccia.

77. Gibbon Mountain sill and Sabino Canyon fault, Sabino Canyon (Sabino Canyon quadrangle). Mylonitized porphyroclastic biotite leucogranite of this sill is exposed at road level northwest of the Sabino Canyon fault, which is represented by red-matrix breccia float. Southeast of the fault, mylonitic Oracle Granite is at road level, but the Gibbon Mountain sill can be seen high on the canyon slopes to the east as a line of cliffs. The form of the Forerange anticline is also apparent from here; the area beyond Thimble Peak is on the north limb.

78. Base of Gibbon Mountain sill in waterfall, Soldier Canyon (Agua Caliente Hill quadrangle). The intrusive lower contact of the leucogranite sill is at the base of the waterfall, and is with Leatherwood granodiorite. Both lithologies are cut by pegmatite dikes; all lithologies are mylonitic.

79. Top of Thimble Peak sill cut by Sabino Canyon fault (Sabino Canyon quadrangle). The upper contact of the sill with Oracle Granite is offset by the Sabino Canyon fault (fig. 55) such that the east side appears to be dropped down. Both lithologies are mylonitic.

80. Interior of Thimble Peak sill, Sabino Canyon trail (Sabino Canyon quadrangle). Coarse porphyroclastic granite predominates. Medium-grained equigranular leocogranite here is less deformed, and locally shows a vague primary banding. Pegmatite dikes here are also little-deformed.

81. Shreve Pass sill near Bear Canyon (Agua Caliente Hill quadrangle). Deformed leucogranite here is near the upper intrusive contact with Oracle Granite. This is the freshest known exposure of the Shreve Pass sill.

82. Catnip Canyon sill in Catnip Canyon at "deadman's curve" on Catalina Highway (Agua Caliente Hill quadrangle). Biotite leucogranite here shows banding thought to be of primary origin. "Igneous unconformities" in this sill suggest upright attitude (fig. 52) but are thought by some to be tectonic. This locality is that from which George Gehrels obtained a U-Pb date on monazite of about 52 Ma.

83. Metadiabase xenolith in leucogranite of inclusion zone, Catalina Highway near Seven Cataracts (Agua Caliente Hill quadrangle). The xenolith (fig. 58) shows a strong refolded foliation, presumably Laramide, but locally preserves relict sub-ophitic texture. Wilderness suite dikelets cut the foliation. Just west is the exposed contact of the inclusion zone with the top of the Catnip Canyon sill (fig. 59).

84. Wilderness granite at Hitchcock picnic area (Mt. Bigelow quadrangle). Massive muscovite-garnet

leucogranite and pegmatite dikes are little-deformed. A monazite U-Pb date reportedly from here gave 44.6±1.5 Ma (Reynolds and others, 1986).

85. Lemmon Rock pegmatite-aplite near Sabino Canyon (Mt. Lemmon quadrangle). Three phases of pegmatite and aplite dikes in Leatherwood granodiorite are found in this large outcrop (fig. 63). Banded aplite with garnet bands forms a phase of intermediate age; this garnet band continues to localities 86 and 17. The youngest phase shows a porphyritic texture in pegmatite.

86. Lemmon Rock pegmatite-aplite near Sabino Canyon (Mt. Lemmon quadrangle). Aplite bands show draping and other directional accumulation patterns (fig. 64). The garnet aplite band of localities 17 and 85 traverses this outcrop.

87. Lemmon Rock pegmatite-aplite below Marshall Gulch in Sabino Canyon (Mt. Lemmon quadrangle). Some pegmatite here is porphyritic (fig. 66); megacrysts of feldspar with graphically intergrown quartz have garnet-mica selvages. The groundmass is muscovite-garnet granite. A xenolith of Oracle Granite forms the east bank.

88. Two generations of pegmatite near Molino Canyon (Agua Caliente Hill quadrangle). Cliff shows an older deformed pegmatite and a younger little-deformed pegmatite cutting it (fig. 63). Both pegmatites contain garnet, and cut mylonitic Oracle Granite and Leatherwood granodiorite.

89. Down-lineation view of pegmatites, Molino Canyon (Agua Caliente Hill quadrangle). The original angular relations of pegmatite dikes and sills are best represented in this view along the lineation of mylonite. Oracle Granite is the host rock. Ptygmatic folds in the pegmatite bodies are displayed looking oblique to lineation across the canyon (fig. 75). The form of the Forerange anticline can be discerned in this area also.

90. Relations of Spencer granite, Lemmon Rock pegmatite-aplite, and Pioneer Formation near Box Camp Trail (Mt. Bigelow quadrangle). A gently dipping screen of Pioneer Formation phyllite separating the two granitic bodies is cut by an apophysis of Spencer granite, which in turn is cut by the structurally underlying Lemmon Rock unit.

91. Leucogranite of the Spencer lens, Catalina Highway (Mt. Bigelow quadrangle). Weakly foliated biotite leucogranite here was dated by George Gehrels at about 55 Ma by U-Pb on monazite.

92. Knagge granite, Knagge mine, Knagge Trail (Mt. Bigelow quadrangle). Biotite leucogranite containing garnet is cut by a mineralized shear zone. A large meta-diabase xenolith is just above.

93. Phases of porphyritic Catalina granite, Canada del Oro area (Oracle quadrangle). Undeformed light-colored porphyry is cut first by darker porphyry, then by rhyolitic porphyry.

94. Cloudburst detachment fault, Oracle Granite, Cloudburst Formation, and rhyolite, Tucson Wash (Mammoth quadrangle). Oracle Granite at wash level is brecciated and is structurally overlain by Cloudburst Formation conglomerate and volcanic rocks dipping into the intervening Cloudburst detachment fault. Rhyolite is intrusive into both packages and is offset by the fault at outcrop scale (fig. 69) but not map scale. At east end of this window through the detachment, intrusive rhyolite is altered to sericite and epidote; see table 5 for rhyolite composition.

95. Refolded fold, Kellogg Mountain (Mt. Bigelow quadrangle). Metadiabase and Pioneer Formation show a tight recumbent fold (fig. 70) with a refolded older lineation. Just downslope, such folding is cut by younger Wilderness granite.

96. Deformed Leatherwood granodiorite dike in Horquilla Limestone, Marble Peak (Mt. Bigelow quadrangle). Horquilla hosts a 1 m Leatherwood granodiorite dike that has been folded and boudinaged (fig. 71). At nearby localities (pl. 2B), however, deformed Horquilla is cut by undeformed Leatherwood dikes.

97. Skarn in Abrigo Formation near Davis Spring Trail (Mt. Bigelow quadrangle). A porphyritic sill of Leatherwood granodiorite has produced cauliflower-shaped diopside masses up to a meter high in the structurally underlying lower (Three C) member of Abrigo Formation, and locally in the intervening middle (Southern Belle) quartzite member.

98. Skarn in Mescal Limestone and Abrigo Formation on Davis Spring Trail (Mt. Bigelow quadrangle). The porphyritic Leatherwood granodiorite sill of locality 97 has produced coarse diopside and finer amphibole both above and below Bolsa Quartzite in Abrigo Formation and Mescal Limestone, respectively.

99. Relations of pegmatite, Pioneer Formation, and Dripping Spring Quartzite in Marshall Gulch (Mt. Lemmon quadrangle). Pegmatite intruded along the gulch forms cliffs, and partially connects the larger pegmatite bodies of Sabino Canyon and Lemmon Rock (pl. 1). Intrusion was along an older fault separating Pioneer Formation to the north from Dripping Spring Quartzite to the south. Pioneer Formation here locally contains coarse garnet.

100. Mylonitic Oracle Granite, west fork Sabino Creek (Mt. Lemmon quadrangle). An older top-to-southwest mylonitic fabric shows an overturned S-surface. The Romero Pass fault on the northeast side of the outcrop juxtaposes Oracle against mylonitic Wilderness granite. Along the trail to the northwest, top-to-southwest fabrics have been overprinted by top-to-northeast fabrics. C-surfaces of the younger fabric are locally strong (fig. 74).

101. Ptygmatic folds in mylonite, Sabino Canyon (Sabino Canyon quadrangle). Mylonitic Oracle Granite between the Gibbon Mountain and Thimble Peak leucogranite sills shows ptygmatic folding of leucogranite dikes (looking oblique to lineation) on the western wall of the canyon viewed from here.

102. Mylonitic Gibbon Mountain sill in Molino basin (Agua Caliente Hill quadrangle). Adjacent domains show C-surfaces dipping alternate directions, either to the southwest (top to southwest) or more steeply northeast than S-surfaces (top to northeast). The base of the sill is exposed on the ridge above.

103. Mylonitic Wilderness granite at Windy Point on Catalina Highway (Agua Caliente Hill quadrangle). Garnet-2 mica granite has been mylonitized with top-to-southwest fabric (C-surface dips southeast, S-surface about horizontal). Thin pegmatite bodies are locally about horizontal also (pl. 1). Mylonitic rock forms a large mullion that grazes the surface of the range at only a few places toward the northeast (pl. 1B) along the lineation. Margins of this mullion as on Rose Peak are visible to the northeast from good vantages at Windy Point. Within the Windy Point area, mylonitic foliation decreases westward down the ridge and eastward along the road in either direction; all these routes lead structurally downward out of the mullion.

104. Mylonitic Wilderness granite and margin of Knagge granite at Leopold Point (Mt. Bigelow quadrangle). This locality marks the northeastern extremity of the Windy Point mylonite zone. Mylonite is best developed on the west side in Wilderness granite. The crags to the east are of less mylonitized Knagge granite.

105. Thin mylonitic zone in Knagge granite at Barnum Rock (Mt. Bigelow quadrangle). Massive Knagge granite is cut by a northeast-dipping mylonite zone a few meters thick. This locality would be within the Windy Point mylonite zone (pl. 1B), but Knagge granite apparently cuts most mylonitization.

106. Structural features at the base of the Windy Point zone, Palisades Trail in Pine Canyon (Mt. Bigelow quadrangle). Slabs of less mylonitic Wilderness granite about 45 m (150 ft) thick alternate with thinner mylonitic zones dipping northeast, as in fig. 76. Visible on a steep wall across Pine Canyon is primary banding in Wilderness Granite cut by mylonite.

107. Sabino Canyon fault and related microgranite dikes, Sabino Canyon road (Sabino Canyon quadrangle). Fault planes with red gouge and down-dip slickensides are locally intruded by thin dikes of biotitic micro-granite, commonly altered to epidote and here containing fluorite. Fault breccia appears toward the northeast along the road.

108. Relation of Romero Pass fault zone to late-stage mylonite, Sabino basin (Sabino Canyon quadrangle). Gouge and breccia of the fault dip south, cutting cataclastic mylonite dipping north.

109. Romero Pass and Sabino Canyon faults, Sabino basin (Sabino Canyon quadrangle). Gouge of Romero Pass zone appears to truncate that of Sabino Canyon fault; however some strands of the Romero Pass zone are under a landslide mass across the valley.

110. Intrusion of Leatherwood granodiorite across Geesaman fault, upper Stratton Canyon north of Marble Peak (Mt. Bigelow quadrangle). Leatherwood intrudes Pioneer Formation on the north wall of the fault.

111. Intrusion of Leatherwood granodiorite across Geesaman fault, Dan Saddle track (Mt. Bigelow quadrangle). The intrusive contact with Pioneer Formation is exposed in the roadbed.

112. Intrusion of Leatherwood granodiorite into Geesaman fault zone, Dan Saddle track (Mt. Lemmon quadrangle). Mylonite mostly after Oracle Granite shows a horizontal fabric. The intrusive contact is not well exposed, but Leatherwood though epidotic is less sheared.

113. Intrusions of felsite in Geesaman fault zone, East Fork Canada del Oro (Mt. Lemmon quadrangle). Alaskite and aplite related to Reef of Rock facies of Catalina granite intrude and thereby obscure the Geesaman fault, which here juxtaposes Oracle Granite and Abrigo Formation.

114. Kellogg fault, Knagge Trail (Mt. Bigelow quadrangle). At this locality, the fault has been intruded by leucogranite of the Knagge pluton. Figure 90 shows the relations visible along the fault to the west.

115. Fault intruded by pegmatites, Sunset Trail (Mt. Lemmon quadrangle). A poorly exposed unconformity between deformed Oracle Granite and Scanlan Conglomerate Member of Pioneer Formation is just above the level of the trail on the northwest side of this valley, whereas southeast of it the same unconformity is over 200 ft higher. The fault plane is obscured by intrusion of abundant pegmatite.

116. Contact of Oracle Granite and Scanlan Conglomerate Member, mine-haul road (Mt. Bigelow quadrangle). Both litholgies show mortar structure, probably as wall-rock deformation on the nearby Geesaman fault, but the juxtaposition appears to be an unconformity rather than a fault.

117. Contact of Oracle Granite and Pioneer Formation, Bat Hill (Mt. Bigelow quadrangle). Basal Pioneer, apparently as an argillaceous pebbly sandstone now stretched and foliated, rests tightly on Oracle. This seems to be an unusual variety of Scanlan Conglomerate Member, and the contact is an unconformity. On the southern (loc. 8) and western flanks of the hill, normal Scanlan reappears. Better-sorted sandstone of the middle Pioneer overlies all the varieties of basal Pioneer. Northward from here, one looks across the core of the Lariat anticline; this outcrop defines its south limb.

118. Fault at Mary West mine (Campo Bonito quadrangle). The upper and middle members of Pioneer Formation are juxtaposed on an old mineralized fault. See text for mine description.

119. Tectonic inclusion in reverse fault (Campo Bonito quadrangle). A large sliver of red argillaceous sandstone, probably of Paleozoic or Apache Group provenance, is caught between the lower member of American Flag Formation and Escabrosa Limestone on one of a family of reverse faults bounding the American Flag.

120. Mogul fault on track near Dodge Wash (Oracle quadrangle). The fault here juxtaposes protoclastic Oracle Granite with quartz veins to the northeast against brecciated Pinal Schist to the southwest. Gouge is only 0.3 m wide.

121. Mogul fault on Control Road near Three C Ranch (Campo Bonito quadrangle). The fault here juxtaposes protoclastic Oracle Granite to the north against brecciated Escabrosa Limestone to the south. Laminated gouge as much as 10 m thick dips south. This interval of the fault may include both the proto- and neo-Mogul faults of this report.

122. The Mogul fault at Cody tunnel, Campo Bonito district (Campo Bonito quadrangle). Oracle Granite and aplite showing mortar texture to the north are juxtaposed against Dripping Spring Quartzite and a diabase dike, across a near-vertical fault just north of the mouth of the tunnel. In the valley below, the fault is offset along a trend now marked by a weathered porphyry dike. Mine description in text.

123. Limb fault near Coronado Camp track (Oracle quadrangle). The fault here juxtaposes shattered Barnes Conglomerate Member of Dripping Spring Quartzite against Pinal Schist metarhyolite. The fault plane is iron-cemented gouge and breccia.

124. Fault repeating Barnes Conglomerate Member of Dripping Spring Quartzite near Oracle Ridge trail (Oracle quadrangle). Barnes is dropped down about 120 feet to the west, thus juxtaposing Dripping Spring and Pioneer Formations. This fault apparently offsets the proto-Mogul fault just north of here.

125. Stratton detachment fault on Control Road (Mt. Bigelow quadrangle). Fault gouge with variable dip separates sheared Oracle Granite below from variably dipping Stratton unit above.

126. Stratton detachment fault in Stratton Canyon (Mt. Bigelow quadrangle). Subhorizontal gouge separates mylonitic quartzite and other Apache Group rocks below, exposed to the southeast, from Stratton volcanics and conglomerate dipping into the fault, exposed to the northwest. The gouge is talc-like below and a red breccia above.

127 Neo-Mogul fault, track in Campo Bonito area (Campo Bonito quadrangle). Oracle Granite forms both walls of a tectonic breccia with chloritic matrix and some open space.

128. Neo-Mogul fault, track near Oracle Ridge Trail (Oracle quadrangle). Tectonic "conglomerate" with rounded Oracle Granite clasts in a chloritic matrix apparently truncates a nearby diabase body on the north wall. e Wash (Oracle quadrangle). A 1-m sericitic gouge zone in Oracle Granite contains a quartz vein.

129. Neo-Mogul fault in Irene Wash (Oracle quadrangle). A 1-m sericitic gouge zone in Oracle Granite contains a quartz vein.

130. Neo-Mogul fault in bed of unnamed wash in San Pedro basin (Campo Bonito quadrangle). Fault breccia with matrix derived from the Stratton unit contains blocks of alaskite and American Flag (?) Formation. Alaskite is present on the northeast wall. This is the southeasternmost known outcrop of the Mogul fault.

131. Cloudburst detachment fault in unnamed wash (Mammoth quadrangle). Epidotic Cloudburst Formation volcanics structurally overlie Oracle Granite in a small window. Brecciated rhyolite is apparently present in both hosts.

132. Turtle fault in Tucson Wash (Mammoth quadrangle). The fault juxtaposes brecciated Oracle Granite to the south against Cloudburst Formation lower member (fig. 97), which locally may represent lahar deposits. The fault plane itself is epidotized, and contains quartz veins with the typical morphology of the Mammoth vein set. Intrusive rhyolite is ubiquitous. South about 100 m, the main Collins segment of the Mammoth vein set is exposed high on the west bank of the wash (cf. fig. 104).

133. Mammoth vein set, intrusive rhyolite, and unconformity of Cloudburst Formation on Oracle Granite, Mammoth mine (Mammoth quadrangle). Conglomerate of the Cloudburst Formation lower member on the western wall is juxtaposed against Oracle Granite overlain by Cloudburst Formation to the east, separated by the complex Mammoth vein. Rhyolite intrudes both lithologies. Mine description in text.

134. San Manuel and other faults in San Manuel open-pit mine (Mammoth quadrangle). The upper (Tucson Wash) member of San Manuel Formation dips eastward into the west-dipping plane of the San Manuel fault, juxtaposing San Manuel Formation against Oracle Granite and San Manuel porphyry (fig. 98). Also exposed in the mine are the East (Mammoth) and West faults. These faults are visible from the mine overlook. Mine description in text.

135. San Manuel fault in Tucson Wash (Mammoth quadrangle). The type section of Tucson Wash Member of San Manuel Formation dips eastward into the San Manuel fault, along which have accumulated large blocks of rhyolite and Cloudburst Formation (fig. 99). Below the fault is Oracle Granite and related rocks. The fault plane here contains black calcite masses.

136. Relations of San Manuel porphyry, San Manuel Formation, and the San Manuel fault, north fork of Mammoth Wash (Mammoth quadrangle). The upper (Tucson Wash) member of San Manuel Formation unconformably overlies San Manuel porphyry type B, and is juxtaposed westward against San Manuel Formation upper member across the San Manuel fault.

137. San Manuel Formation, Cloudburst Formation, and Oracle Granite in steep gully along San Manuel fault (Mammoth quadrangle). The fault juxtaposes Oracle Granite to the east against Cloudburst Formation upper member unconformably overlain by upper (Tucson Wash) member of San Manuel Formation.

138. Rhyolitic tuff in Cloudburst Formation, Tucson Wash (Mammoth quadrangle). Light-colored tuff a few meters thick is apparently air-fall at the base and ash-flow, perhaps welded in part, toward the top. The tuff is enclosed in conglomerate of Cloudburst Formation lower member.

139. American Flag mine (Campo Bonito quadrangle). Mine description in appendix.

140. Bearcat mine (Campo Bonito quadrangle). Mine description in appendix.

141. Magrugador mine (Campo Bonito quadrangle). Mine description in appendix.

142. North Cross workings (Oracle quadrangle). Description of workings in appendix.

143. Workings on Oracle Hill (Oracle quadrangle). Description of workings in appendix.

144. Dead Bull mine (Campo Bonito quadrangle). Mine description in appendix.

145. Altered American Flag Formation (Campo Bonito quadrangle). Upper member of American Flag has been altered to propylitic, then quartz sericite assemblages along northeast-trending quartz-chlorite veins and faults.

146. Workings along Geesaman Creek (Mt. Bigelow quadrangle). Description of workings in appendix.

147. Bluff mine group (Mt. Bigelow quadrangle). Mine descriptions in appendix.

148. Geesaman mine group of Peterson and Creasey (1943) (Mt. Bigelow quadrangle). Mine descriptions in appendix. Note that location differs from topographic map.

149. Daily mine group of Peterson and Creasey (1943) (Mt. Bigelow quadrangle). Mine descriptions in appendix. Note that location differs from topographic map.

150. Unnamed workings, Atchley Canyon (Mt. Bigelow quadrangle). Description of workings in appendix.

151. Zimmerman mine (Mt. Lemmon quadrangle). Mine description in appendix.

152. Mammoth-type rhyolite veinlets in rhyolite (Mammoth quadrangle). Anastomosing quartz veinlets with dark margins in rhyolite cutting the unconformity between Oracle Granite and overlying Cloudburst Formation .

153. Mammoth-like veinlets in Cloudburst Formation (Mammoth quadrangle). Quartz and calcite veins in conglomeratic Cloudburst Formation near the trace of the Turtle fault.

REFERENCES CITED

Anderson, S. R., 1987, Cenozoic stratigraphy and geologic history of the Tucson Basin, Pima County, Arizona: U. S. Geological Survey Water Resources Investigations Report 87-4190, 20p.

Anderson, J. L., Barth, A. P., and Young, E. D., 1988, Mid-crustal Cretaceous roots of Cordilleran metamorphic core complexes: Geology, v. 16, p. 366–369.

Banerjee, A. K., 1957, Structure and petrology of the Oracle Granite, Pinal County, Arizona: Tucson, University of Arizona, unpub. Ph.D. thesis, 112 p.

Banks, N. G., 1974, Reconnaissance geologic map of the Mount Lemmon quadrangle, Arizona, U. S. Geological Survey Miscellaneous Field Studies Map MF 747, scale 1:62,500.

———, 1980, Geology of a zone of metamorphic core complexes in southeastern Arizona, *in* Crittenden, M. D., Coney, P. J., and Davis, G. H., eds. Cordilleran metamorphic core complexes: Geological Society of America Memoir 153, p. 177–215.

Bateman, P. C., and Erickson, M. P., 1944, Maudina and Morning Star tungsten mines, Pinal County, Arizona: U. S. Geological Survey Open-file maps and text.

Bideaux, R. A., 1980, Tiger, Arizona: Mineralogical Record, v. 11, no. 3, p. 155–181.

Bilodeau, W. L., 1982, Tectonic models for Early Cretaceous rifting in southeastern Arizona: Geology, v. 10, p. 466–470.

Bilodeau, W. L., 1987,: Arizona Geological Society Digest, v. 18, p. 229–256.

Blakey, R. C., and Knepp, Rex, 1989, Pennsylvanian and Permian geology of Arizona, *in* Jenney, J. P., and Reynolds, S. J., eds., Geologic evolution of Arizona: Arizona Geological Society Digest, v. 19, p. 313–348.

Bond, G. C., Nickeson, P. A., and Kominz, M. A., 1984, Breakup of a supercontinent between 625 Ma and 555 Ma; new evidence and implications for continental histories: Earth and Planetary Science Letters, v. 70, p. 325–345.

Braun, E. R., 1969, Geology and ore deposits of the Marble Peak area, Santa Catalina Mountaings, Pima County, Arizona: Tucson, University of Arizona, unpub. M. S. thesis, 79 p.

Bromfield, C. S., 1950, Geology of the Maudina mine area, northern Santa Catalina Mountains, Pinal County, Arizona: Tucson, University of Arizona, unpub. M. S. thesis, 63 p.

———, 1952, Some geologic features of the Santa Catalina Mountains: Arizona Geological Society and Geological Society of America, Guidebook Southern Arizona, p. 51–55.

Bryant, D. L., 1968, Diagnostic characteristics of the Paleozoic formations of southeastern Arizona, *in* Titley, S. R., ed., Southern Arizona Guidebook III: Arizona Geological Society, p. 33–47.

Bykerk-Kauffman, Ann, 1986, Multiple episodes of ductile deformation within the lower plate of the Santa Catalina metamorphic core complex, *in* Beatty, B., and Wilkinson, P. A. K., eds., Frontiers in geology and ore deposits of Arizona and the Southwest: Arizona Geological Society Digest, v. 16, p. 460–463.

———, 1990, Structural evolution of the northeastern Santa Catalina Mountains, Arizona: a glimpse of the pre-extension history of the Catalina complex: Tucson, University of Arizona, unpub. Ph.D. dissertation, 195 p.

Catanzaro, E. J., and Kulp, J. L., 1964, Discordant zircons from the Little Belt (Montana), Beartooth (Montana), and Santa Catalina (Arizona) Mountains: Geochimica et Cosmochimica Acta, v. 28, p. 87–124.

Cobbald, P. R., and Quinquis, H., 1980, Development of sheath folds in shear regimes: Journal of Structural Geology, v. 2, p. 119–126.

Coney, P. J., 1980, Cordilleran metamorphic core complexes: an overview, *in* Crittenden, M. D., Coney, P. J., and Davis, G. H., eds., Cordilleran metamorphic core complexes: Geological Society of America Memoir 153, p. 7–34.

——— and Reynolds, S. J., 1980, Cordilleran metamorphic core

complexes and their uranium favorability (final report and appendices): U. S. Department of Energy, Open-file Report GJBX-258(80), 627p.

Corbitt, L. L., and Woodward, L. A., 1981, Late Precambrian tillite, Florida Mountains, southwestern New Mexico *in* Hambrey, M. J., and Harland, W. B., eds., Earth's pre-Pleistocene glacial record: Cambridge Univ. Press, p. 749–750.

Creasey, S. C., 1950, Geology of the St. Anthony (Mammoth) area, Pinal County, Arizona, in Arizona zinc and lead deposits, part 1, chapter 6: Arizona Bureau of Mines Bulletin 156, p. 63–84.

———, 1965, Geology of the San Manuel area, Pinal County, Arizona: U.S. Geological Survey Professional Paper 471, 64 p.

———, 1967, General geology of the Mammoth Quadrangle, Pinal County, Arizona: U. S. Geological Survey Bulletin 1218, 94 p.

Creasey, S. C., and Theodore, T. G., 1975, Preliminary reconnaissance geologic map of the Bellota Ranch quadrangle, Pima County, Arizona: U. S. Geological Survey Open-file Report 75–295.

Dale, V. B., 1959, Tungsten deposits of Yuma, Maricopa, Pinal, and Graham counties, Arizona: U. S. Bureau of Mines Report of Investigations 5516, 68 p.

Damon, P. E., Erickson, R. C., and Livingston, D. E., 1963, K-Ar dating of basin and range uplift, Catalina Mountains, Arizona: Nuclear Geophysics, National Academy of Sciences, NRC Publication 1075, p. 113–121.

Davidson, E. S., 1973, Geohydrology and water resources of the Tucson basin, Arizona: U. S. Geological Survey Water-supply Paper 1939-E, 81 p.

Davis, G. A., Lister, G. S., and Reynolds, S. J., 1986, Structural evolution of the Whipple and South Mountains shear zones, southwestern United States: Geology, v. 14, p. 7–10.

Davis, G. A., Anderson, J. L., Martin, D. L., Krummenacher, Daniel, Frost, E. G., and Armstrong, R. L., 1982, Geologic and geochronologic relations in the lower plate of the Whipple detachment fault, Whipple Mountains, southwestern California: a progress report, *in* Frost, E. G., and Martin, D. L., eds., Mesozoic-Cenozoic tectonic evolution of the Colorado River region, California, Arizona, and Nevada: Cordilleran Publishers, San Diego, p. 408–432.

Davis, G. H.,1979, Laramide folding and faulting in southeastern Arizona: Americal Journal of Science, v. 279, p. 543–569.

———, 1980, Structural characteristics of metamorphic core complexes, southern Arizona, *in* Crittenden, M. D., Coney, P. J., and Davis, G. H., eds., Cordilleran metamorphic core complexes: Geological Society of America Memoir 153, p. 35–77.

——— and Coney, P. J., 1979, Geologic development of Cordilleran metamorphic core complexes: Geology, v. 7, p. 120–124.

Davy, Philippe, Guerin, Gwenael, and Brun, J.-P., 1989, Thermal constraints on the tectonic evolution of a metamorphic core complex (Santa Catalina Mountains, Arizona): Earth and Planetary Science Letters, v. 94, p. 425–440.

Dempsey, W. J., and Hill, M. E., 1963, Aeromagnetic map of the Mammoth quadrangle, Pinal and Pima counties, Arizona: U. S. Geological Survey Geophysical Investigations Map GP 419 (1:62,500).

Dickinson, W. R., 1987, General geologic map of Catalina core complex and San Pedro trough: Arizona Geological Survey Miscellaneous Map MM-87-A (1:62,500).

———, 1991, Tectonic setting of faulted Tertiary strata associated with the Catalina core complex in southern Arizona: Geological Society of America Special Paper 264, 106 p.

———, 1993, Summary geologic map of the Black Hills near Mammoth, Pinal County, Arizona: Arizona Geological Survey Contributed Map CM 93B (1:24,000).

Dickinson, W. R., and Shafiqullah, M., 1989, K-Ar and F-T ages for syntectonic mid-Tertiary volcanosedimentary sequences associated with the Catalina core complex and San Pedro trough in southern Arizona: Isochron/West, no. 52, p. 15–27.

Dickinson, W. R., Kluth, M. A., and Swift, P. J., 1986, The Bisbee Basin and its bearing on late Mesozoic paleogeographic and paleotectonic relations between the Cordilleran and Caribbean regions, *in* Abbott, P. L., ed., Cretaceous stratigraphy, western North America: Society of Economic Paleontologists and Mineralogists, Pacific Section Book 46, p. 51–62.

Dickinson, W. R., Force, E. R., and Hagstrum, J. T., 1996, Tilting history of the San Manuel-Kalamazoo porphyry system, southeastern Arizona—a reply: Economic Geology, v. 90, p. 2096–2098.

Eberly, L. D., and Stanley, T. B., Jr., 1978, Cenozoic stratigraphy and geologic history of southwestern Arizona: Geological Society of America Bulletin, v. 89, p. 921–940.

Epis, R. C., and Gilbert, C. M., 1957, Early Paleozoic strata in southeastern Arizona: American Association Petroleum Geologists Bulletin, v. 41, p. 2223–2242.

Force, E. R., and Cox, L. J., 1992, Structural context of mid-Tertiary mineralization in the Mammoth and San Manuel districts, southeastern Arizona: U. S. Geological Survey Bulletin 2042C, 28 p.

Force, E. R., and Dickinson, W. R., 1994, Tucson Wash—an introduction to new work in the San Manuel and Mammoth districts, Pinal County, Arizona: U. S. Geological Survey Circular 1103B, p. 103–115 (and Arizona Geological Society field trip guidebook, Spring 1993, 24 p).

Force, E. R., Dickinson, W. R., and Hagstrum, J. T., 1995, Tilting history of the San Manuel-Kalamazoo porphyry, southeastern Arizona: Economic Geology, v. 90, p. 67–80.

Force, E. R., in press, Laramide alteration of Proterozoic diabase: a likely contributor of copper to porphyry systems in the Dripping Spring Mountains area, southeastern Arizona: Economic Geology.

Frost, E. G., and Heidrick, T. L., 1966, Three-dimensional structural geometries of the Colorado River extensional terranc and their regional exploration implications: Society of Economic Geologists Guidebook 25, p. 5–65.

Goodlin, T. C., and Mark, R. A., 1987, Tectonic implications of stratigraphy and structure of Cretaceous rocks and overlying mid-Tertiary cover in Hot Springs Canyon, Galiuro Mountains, southeastern Arizona, *in* Dickinson, W. R., and Klute, M. A., eds., Mesozoic rocks in southern Arizona and adjacent areas: Arizona Geological Society Digest v. 18, p. 177–188.

Granger, H. C., and Raup. R. B., 1964, Stratigraphy of the Dripping Spring Quartzite, southeastern Arizona: U. S. Geological Survey Bulletin 1168, 119 p.

Guerin, Gwenael, Brun, J.-P., and Van Den Driessche, Jean, 1990, Kinematics of pre-Miocene ductile deformation in the Santa Catalina core complex and adjacent regions: Tectonics, v. 9, p. 1305–1326.

Guilbert, J. M., and Lowell, J. D., 1996, Tilting history of the San Manuel-Kalamazoo porphyry system, southeastern Arizona—a discussion: Economic Geology, v. 90, p. 2094–2096.

Hansen, J. B., 1983, Style of deformation of upper-plate rocks of the San Manuel-Camp Grant low-angle normal fault system, Black Hills, Pinal County, Arizona: Tucson, University of Arizona, unpub. M. S. thesis, 164 p.

Hanson, H. S., 1966, Petrography and structure of the Leatherwood Quartz Diorite, Santa Catalina Mountains, Pima County, Arizona: Tucson, University of Arizona, unpub. Ph.D. thesis, 104 p.

Harrison, A. E., 1972, The Santa Catalinas—a description and history: unpub. manuscript, U. S. Forest Service, (Sabino Canyon Visitor Center, Tucson), 195 p.

Haxel, G. B., Tosdal, R. M., May, D. J, and Wright, J. E., 1984, Latest Cretaceous and early Tertiary orogenesis in south-central Arizona: thrust faulting, regional metamorphism, and granitic intrusion: Geological Society of America Bulletin, v. 95, p. 631–653.

Hayes, P. T., 1972, Stratigraphic nomenclature of Cambrian and Lower

Ordovician rocks of easternmost southern Arizona and adjacent westernmost New Mexico: U. S. Geological Survey Bulletin, 1372-B, 21 p.
Hayes, P. T., and Landis, E. R., 1965, Paleozoic stratigraphy of the southern part of the Mule Mountains, Arizona: U. S. Geological Survey Bulletin, 1201-F, 43 p.
Heindl, L. A., 1963, Cenozoic geology in the Mammoth area, Pinal County, Arizona: U. S. Geological Survey Bulletin, 1141-E, 41 p.
Heylmun, E. B., 1989, Santa Catalina Mountains, Arizona: California Mining Journal, v. 69, p. 11–15.
Hinkle, M. E., Kilburn, J. E., Eppinger, R. G. III, and Tripp, R. B., 1981, Geochemical maps of the Pusch Ridge Wilderness Area, Pima County, Arizona: U. S. Geological Survey Map MF 1356A.
Hinkle, M. E., and Ryan, G. S., 1982, Mineral resource potential of the Pusch Ridge Wilderness Area, Pima County, Arizona: U. S. Geological Survey Map MF 1356B.
Howard, K. A., 1991, Intrusion of horizontal dikes; tectonic significance of middle Proterozoic diabase sheets widespread in the upper crust of the southwestern United States: Journal of Geophysical Research, v. 96, p. 12461–12478.
Janecke, S. A., 1986, Structural geology and tectonic history of the Geesaman Wash area, Santa Catalina Mountains, Arizona: Tucson, University of Arizona, unpub. M. S. thesis, 151 p., also available as Arizona Bureau of Geology and Mineral Technology Miscellaneous Map MM 86A.
———, 1987, Mesozoic and early Tertiary structural history of the Geesaman Wash area, northeastern Catalina-Rincon metamorphic core complex, *in* Dickinson, W. R., and Klute, M. A., eds., Mesozoic geology of southern Arizona and adjacent areas: Arizona Geological Society Digest, v. 18, p. 189–201.
John, B. E., and Mukasa, S. B., 1990, Footwall rocks to the mid-Tertiary Chemehuevi detachment fault; a window into the middle crust in the southern Cordillera: Journal of Geophysical Research, v. 95, p. 463–465.
Johnson, M. G., 1972, Placer gold deposits of Arizona: U. S. Geological Survey Bulletin 1355, 103 p.
Karlstrom, K. E., and Bowring, S. A., 1988, Early Proterozoic assembly of tectonostratigraphic terranes in southwestern North America: Journal of Geology, v. 96, p. 561–576.
Keith, S. B., 1974, Index of mining properties in Pima County, Arizona: Arizona Bureau of Mines Bulletin 189, 156 p.
Keith, S. B., Reynolds, S. J., Damon, P. E., Shafiqullah, Muhammad, Livingston, D. E., and Pushkar, P. D., 1980, Evidence for multiple intrusion and deformation within the Santa Catalina-Rincon-Tortolita crystalline complex, southeastern Arizona, *in* Crittenden, M. D., Coney, P. J., and Davis, G. H., eds., Cordilleran metamorphic core complexes: Geological Society of America Memoir 153, p. 217–267.
Kluth, C. F., 1982, The geology and mid-Mesozoic tectonics of the northern Canelo Hills, Santa Cruz County, Arizona: Tucson, University of Arizona, unpub. Ph.D. dissertation, 245 p.
Krantz, R. W., 1989, Laramide structures of Arizona, *in* Jenney, J. P., and Reynolds, S. J., eds., Geologic evolution of Arizona: Arizona Geological Society Digest, v. 19, p. 463–483.
Krieger, M. H., 1968, Stratigraphic relations of the Troy Quartzite (younger Precambrian) and the Cambrian formations in southeastern Arizona, *in* Titley, S. R., ed., Southern Arizona Guidebook III: Arizona Geological Society, p. 23–32.
Kruger, J. M., and Johnson, R. A., 1994, Raft model of crustal extension: evidence from seismic reflection data in southeast Arizona: Geology, v. 22, p. 351–354.
Laughlin, A. W., 1960, Petrography of the Molino Basin area of the Santa Catalina Mountains, Arizona: Tucson, University of Arizona, unpub. M. S. thesis, 53 p.
Lipman, P. W., 1993, Geologic map and cross sections of the Tucson Mountains caldera, Arizona: U. S. Geological Survey Map I-2205, 1:24,000.
Long, K. R., 1992, Preliminary descriptive model for detachment fault-related mineralization: U. S. Geological Survey Bulletin 2004, p. 52–56.
Lowell, J. D., 1968, Geology of the Kalamazoo orebody, San Manuel district, Arizona: Economic Geology, v. 63, p. 645–654.
Lowell, J. D. and Guilbert, J. M., 1970, Lateral and vertical alteration-mineralization zoning in porphyry ore deposits: Economic Geology, v. 65, p. 373–408.
Ludden, R. W., 1950, Geology of the Campo Bonito area, Oracle, Arizona: Tucson, University of Arizona, unpub. M. S. thesis, 52 p.
Matter, Phillip, 1969, Petrochemical variations across some Arizona pegmatites and their enclosing rocks: Tucson, University of Arizona, unpub. Ph.D. dissertation, 173 p.
Maynard, J. B., 1991, Iron—Syngenetic deposition controlled by the evolving ocean-atmosphere system, *in* Force, E. R., Eidel, J. J., and Maynard, J. B., eds., Sedimentary and diagenetic mineral deposits—a basin analysis approach to exploration: Reviews in Economic Geology, v. 5, p. 141–146.
McCullough, E. J., 1963, A structural study of the Pusch Ridge-Romero Canyon area, Santa Catalina Mountains, Arizona: Tucson, University of Arizona, unpub. Ph.D. thesis, 67 p.
Middleton, L. T., 1989, Cambrian and Ordovician depositional systems in Arizona, *in* Jenney, J. P., and Reynolds, S. J., eds., Geologic evolution of Arizona: Arizona Geological Society Digest, v. 19, p. 273–286.
Myers, S. C., and Beck, S. L., 1994, Evidence for a local crustal root beneath the Santa Catalina core complex, Arizona: Geology, v. 22, p. 223–226.
Naruk, S. J., 1987, Kinematic significance of mylonitic foliation: Tucson, University of Arizona, unpub. Ph.D. dissertation, 72 p.
——— and Bykerk-Kauffman, Ann, 1990, Late Cretaceous and Tertiary deformation of the Santa Catalina metamorphic core complex, Arizona: Arizona Geological Survey Special Paper 7, p. 41–50.
Nockolds, S. R., 1954, Average chemical compositions of some igneous rocks: Geological Society of America Bulletin, v. 65, p. 1007–1032.
Palais, D. G., and Peacock, S. M., 1990, Metamorphic and stratigraphic constraints on the evolution of the Santa Catalina Mountains metamorphic core complex, Arizona: Journal of Geophysical Research, v. 95, p. 501–507.
Pashley, E. F., 1966, Structure and stratigraphy of central, northern, and eastern parts of the Tucson basin, Arizona: Tucson, University of Arizona, unpub. Ph. D. thesis, 273 p.
Peirce, F. L., 1958, Structure and petrography of part of the Santa Catalina Mountains: Tucson, University of Arizona, unpub. Ph.D. thesis, 86 p.
Peirce, H. W., 1979, The Mississippian and Pennsylvanian (Carboniferous) Systems in the United States—Arizona: U. S. Geological Survey Professional Paper 1110-Z, 20 p.
Pelletier, J. D., 1957, Geology of the San Manuel mine: Mining Engineering, v. 7, p. 760.
Pelletier, J. D., and Creasey, S. C., 1965, Ore deposits, *in* Geology of the San Manuel area, Pinal County, Arizona: U.S. Geological Survey Professional Paper 471, p. 29–59.
Peterson, N. P., 1938, Geology and ore deposits of the Mammoth mining camp area, Pinal County, Arizona: Arizona Bureau of Mines, Bulletin 144, 63 p.
Peterson, N. P., and Creasey, S. C.,1943, Some copper deposits in the Old Hat mining district, Pima Couty, Arizona: U. S. Geological Survey Open-file Report, 13 p.
Peterson, R. C., 1968, A structural study of the east end of the Catalina forerange, Pima County, Arizona: Tucson, University of Arizona, unpub. Ph. D. thesis, 105 p.
Pilkington, H. D., 1962, Structure and petrology of a part of the east flank of the Santa Catalina Mountains, Pima County, Arizona: Tucson, Univ. Arizona, unpub. Ph. D. thesis, 120 p.
Rehrig, W. A., 1982, Metamorphic core complexes of the southwestern United States—an updated analysis, *in* Frost, E. G., and Martin, D. L., eds., Mesozoic-Cenozoic tectonic evolution of the Colorado

River region, California, Arizona, and Nevada: Cordilleran Publishers, San Diego, p. 551–560.

Reynolds, S. J., and Lister, G. S., 1990, Folding of mylonitic zones in Cordilleran metamorphic core complexes—evidence from near the mylonitic front: Geology, v. 18, p. 216–219.

Reynolds, S. J., Florence, F. P., Welty, J. W., Roddy, M. S., Currier, D. A., Anderson, A. V., and Keith, S. B., 1986, Compilation of radiometric age determinations in Arizona: Arizona Geological Survey Bulletin 197, 258 p.

Roddy, M. S., Reynolds, S. J., Smith, B. M. and Ruiz, J., 1988, K-metasomatism and detachment-related mineralization, Harcuvar Mountains, Arizona: Geological Society of America Bulletin, v. 100, p. 1627–1639.

Ross, C. A., 1978, Pennsylvanian and Early Permian depositional framework, southeastern Arizona, *in* Callender, J. F., Wilt, J. C., Clemons, R. E., and James, H. L, eds., Land of Cochise, southeastern Arizona: New Mexico Geological Society 29th Annual Field Conference Guidebook, p. 193–200.

Ruf, A. S., Naruk, S. J., Butler, R. F., and Calderone, G. J., 1988, Strain and magnetic fabric in the Santa Catalina and Pinaleno mountains core complex mylonite zones, Arizona: Tectonics, v. 7, p. 235–248.

Ryan, G. S., 1982, Mine and prospect map of the Pusch Ridge Wilderness Area, Pima County, Arizona: U. S. Geological Survey Map MF-1356-C.

Schumaker, D., 1978, Devonian stratigraphy and correlations in southeastern Arizona *in* Callender, J. F., Wilt, J. C., Clemons, R. E., and James, H. L., ed.s, Land of Cochise, southeastern Arizona: New Mexico Geological Society 29th Annual Field Conference Guidebook, p. 175–181.

Schwartz, G. M., 1953, Geology of the San Manuel copper deposit, Arizona: U.S. Geological Survey Professional Paper 256, 63 p.

Shafiqullah, Muhammad, Pierce, H. W., and Damon, P. E., 1988, Illite K-Ar dating and the Catalina detachment fault: Geological Society of America Abstracts with Programs v. 20, #7, p. A230–231.

Shakel, D. W., 1974, The geology of layered gneisses in part of the Santa Catalina forerange, Pima County, Arizona: Tucson, University of Arizona, unpub. M. S. thesis, 233 p.

———, 1978, Supplemental road log no. 2, Santa Catalina Mountains via Catalina Highway, *in* Callender, J. F., Wilt, J. C., Clemons, R. E., and James, H. L., Land of Cochise, southeastern Arizona: New Mexico Geological Society 29th Annual Field Conference Guidebook, p. 105–111.

Shakel, D. W., Silver, L. T., and Damon, P. E., 1977, Observations on the history of the gneissic core complex, Santa Catalina Mts., southern Arizona (abs.): Geological Society of America Abstracts with Programs, v. 9, p. 1169–1170.

Sherwonit, W. E., 1974, A petrographic study of the Catalina Gneiss in the forerange of the Santa Catalina Mountains, Arizona: Tucson, University of Arizona, unpub. M. S. thesis, 165 p.

Shride, A. F., 1967, Younger Precambrian geology in southern Arizona: U. S. Geological Survey Professional Paper 566, 89 p.

Silver, L. T., 1978, Precambrian formations and Precambrian history in Cochise County, southeastern Arizona, *in* Callender, J. F., Wilt, J. C., Clemons, R. E., and James, H. L., eds., Land of Cochise, southeastern Arizona: New Mexico Geological Society 29th Field Conference Guidebook, p. 157–163.

Sisson, T. W., Moore, J. G., and Czamanske, G. K., 1988, Cumulates, flame rock, and banded aplites at the top of a typical Sierran granodiorite, Emerald Lake, California: EOS, v. 69, p. 1490.

Spencer, J. E., 1984, Role of tectonic denudation in warping and uplift of low-angle normal faults: Geology, v. 12, p. 95–98.

Spencer, J. E., and Reynolds, S. J., 1989, Middle Tertiary tectonics of Arizona and adjacent areas, *in* Jenney, J. P., and Reynolds, S. J., eds., Geologic evolution of Arizona: Arizona Geological Society Digest, v. 19, p. 539–574.

Spencer, J. E., and Welty, J. W., 1986, Possible controls of base- and precious metal mineralization associated with Tertiary detachment faults in the lower Colorado River trough, Arizona and California: Geology, v. 14, p. 195–198.

Stewart, J. H., 1972, Initial deposits in the Cordilleran geosyncline; evidence of a late Precambrian (850 m.y.) continental separation: Geological Society of America Bulletin, v. 83, p. 1345–1360.

Steele, H. J., and Rubly, G. R., 1947, San Manuel prospect: American Institute of Mining and Metallurgical Engineers, Technical Publication 2255, 12 p.

Stoyanow, A. A., 1936, Correlation of Arizona Paleozoic formations: Geological Society of America Bulletin, v. 47, p. 459–540.

Suemnicht, G. A., 1977, The geology of the Canada del Oro headwaters, Santa Catalina Mountains, Arizona: Tucson, University of Arizona, unpub. M. S. thesis, 108 p., also available as Arizona Bureau of Geology and Mineral Technology Miscellaneous Map 77A.

Teichert, Curt, and Schopf, J. M., 1958, A middle or lower Devonian psilophyte flora from Arizona and its paleogeographic significance: Journal of Geology, v. 66, p. 208–217.

Thomas, L. A., 1966, Geology of the San Manuel ore body, *in* Titley, S. R. and Hicks, C. L., eds. Geology of the porphyry copper deposits, southwestern North America: Tucson, University of Arizona Press, p. 133–142.

Voelger, Klaus, 1953, Cenozoic deposits in the southern foothills of the Santa Catalina Mountains near Tucson, Arizona: Tucson, University of Arizona, unpub. M. S. thesis, 104 p.

Waag, C. J., 1968, Structural geology of the Mt. Bigelow-Bear Wallow-Mt. Lemmon area, Santa Catalina Mountains, Arizona: Tucson, University of Arizona, unpub. Ph. D. dissertation, 133 p.

Wallace, R. M., 1954, Structures at the northern end of the Santa Catalina Mountains, Arizona: Tucson, University of Arizona, unpub. Ph. D. thesis, 45 p.

Weibel, W. L., 1981, Depositional history and geology of the Cloudburst Formation near Mammoth, Arizona: Tucson, University of Arizona, unpub. M. S. thesis, 81 p.

Willden, Ronald, 1964, Geology of the Christmas quadrangle, Gila and Pinal Counties, Arizona: U. S. Geological Survey Bulletin 1161-E, 64 p.

Wilson, E. D., 1941, Tungsten deposits of Arizona: Arizona Bureau of Mines Bulletin, 148, 54 p.

Wood, M. M., 1963, Metamorphic effects of the Leatherwood Quartz Diorite, Santa Catalina Mountains, Pima County, Arizona: Tucson, University of Arizona, unpub. M. S. thesis, 68 p.

Wright, J. E., and Haxel, Gordon, 1982, A garnet-two-mica granite, Coyote Mountains, southern Arizona: geologic setting, uranium-lead isotopic systematics of zircon, and nature of the granite source region: Geological Society of America Bulletin, v. 93, p. 1176–1188.

Wrucke, C. T., 1989, The Middle Proterozoic Apache Group, Troy Quartzite, and associated diabase of Arizona, *in* Jenney, J. P. and Reynolds, S. J., eds., Geological evolution of Arizona: Arizona Geological Society Digest, v. 17, p. 239–258.

Young, D. P., 1988, The history of deformation and fluid phenomena in the top of the Wilderness Suite, Santa Catalina Mountains, Pima County, Arizona: Tucson, University of Arizona, unpub. M. S. thesis, 144 p.

Zartman, R. E., 1984, Lead, strontium, and neodymium isotopic characterization of mineral deposits relative to their geologic environments: Proceedings of the 27th International Geological Congress, v. 12, p. 83–106.

Zartman, R. E., and Doe, B. R., 1981, Plumbotectonics—the model: Tectonophysics, v. 75, p. 135–162.